About Island Press

Island Press is the only nonprofit organization in the United States whose principal purpose is the publication of books on environmental issues and natural resource management. We provide solutions-oriented information to professionals, public officials, business and community leaders, and concerned citizens who are shaping responses to environmental problems.

In 1994, Island Press celebrated its tenth anniversary as the leading provider of timely and practical books that take a multidisciplinary approach to critical environmental concerns. Our growing list of titles reflects our commitment to bringing the best of an expanding body of literature to the environmental community throughout North America and the world.

Support for Island Press is provided by Apple Computer, Inc., The Bullitt Foundation, The Geraldine R. Dodge Foundation, The Energy Foundation, The Ford Foundation, The W. Alton Jones Foundation, The Lyndhurst Foundation, The John D. and Catherine T. MacArthur Foundation, The Andrew W. Mellon Foundation, The Joyce Mertz-Gilmore Foundation, The National Fish and Wildlife Foundation, The Pew Charitable Trusts, The Pew Global Stewardship Initiative, The Rockefeller Philanthropic Collaborative, Inc., and individual donors.

About the Safe Energy Communication Council

The Safe Energy Communication Council (SECC) is a national, nonprofit coalition of 11 environmental and public interest media groups. Since 1980 SECC has educated the public, the press, and decisionmakers about the ability of energy efficiency and renewable energy to provide a larger share of our nation's energy needs and has raised public awareness of the economic and environmental liabilities of nuclear power. SECC provides local, state, and national organizations with technical assistance through media skills training and outreach strategies.

Reinventing
Electric Utilities

Reinventing Electric Utilities

Competition, Citizen Action, and Clean Power

Ed Smeloff and
Peter Asmus

Foreword by
Amory Lovins

ISLAND PRESS

Washington, D.C. ■ Covelo, California

Library of Congress Cataloging-in-Publication Data

Smeloff, Ed.
    Reinventing electric utilities: competition, citizen action, and
clean power / Ed Smeloff and Peter Asmus; foreword by Amory
Lovins
        p.   cm.
    Includes bibliographical references and index.
    ISBN 1-55963-454-5 (cloth).—ISBN 1-55963-455-3
    1. Electric utilities—United States. 2. Electric power-plants—
Decentralization—United States. 4. Electric power-plants—Environmental
aspects—United States. 5. Energy conservation—United States.
6. Renewable energy sources—United States. 7. Sacramento Municipal
Utility District (Calif.) I. Asmus, Peter. II. Title.
HD9685.U5S55 1997
333.79'32'0973—dc20   96-31799
                                                                 CIP

Printed on recycled, acid-free paper ♽

Manufactured in the United States of America

10 9 8 7 6 5 4 3 2 1

Contents

Acronyms and Abbreviations

AWEA	American Wind Energy Association
BPA	Bonneville Power Administration
BRPU	Biennial Resource Plan Update
Btu	British thermal unit
CEC	California Energy Commission
CEERT	Center for Energy Efficiency and Renewable Technologies
CFC	Chlorofluorocarbons
CLECA	California Large Energy Consumers Association
CLF	Conservation Law Foundation
CPUC	California Public Utilities Commission
CTC	Competition Transition Charge
DOE	Department of Energy
DSI	Direct Services Industries
DSM	Demand-side management
EDF	Environmental Defense Fund
EPRI	Electric Power Research Institute
FCC	Federal Communications Commission
FERC	Federal Energy Regulatory Commission
GAO	General Accounting Office
GWh	Gigawatt hour
IEP	Independent Energy Producers
INPO	Institute of Nuclear Power Operations

IRP	Integrated Resource Planning
ISFSI	Independent Spent Fuel Storage Installation
ISO	Independent System Operator
JPA	Joint Power Agency
kW	Kilowatt(s)
kWh	Kilowatt hour(s)
MOU	Memorandum of Understanding
MW	Megawatt(s)
NCAC	Northwest Conservation Act Coalition
NEES	New England Electric Service
NEPOOL	New England Power Pool
NOPR	Notice of Public Rulemaking
NRC	Nuclear Regulatory Commission
NRDC	Natural Resources Defense Council
NSP	Northern States Power
O&M	Operations and Maintenance
PBR	Performance Based Ratemaking
PGE	Portland General Electric
PG&E	Pacific Gas and Electric
PUC	Public Utilities Commission
PUCT	Public Utilities Commission of Texas
PURPA	Public Utilities Regulatory Policy Act
PV	Photovoltaic
QF	Qualifying facility
R&D	Research and Development
RFP	Request for Proposals
RNP	Renewables Northwest Project
SCE	Southern California Edison
SDG&E	San Diego Gas and Electric
SMUD	Sacramento Municipal Utility District
SONGS	San Onofre Nuclear Generating Station
T&D	Transmission and Distribution
TURN	Toward Utility Rate Normalization
TVA	Tennessee Valley Authority
WPPSS	Washington Public Power Supply System

Foreword

Over the past two years there has been much sound and fury over proposals to restructure the electric services industry in the United States. The regulatory turbulence has created a flurry of supercharged news reports in the trade, professional, and public press predicting that consumers will soon be able to pick and choose who will sell them electrons. The perception that a new electric order is emerging, where customers will be pushing and shoving to get the cheapest electrons, has created a boutique industry of consultants who are advising utility executives about how to prepare for a more competitive future.

The advice of many of these consultants has been to treat employees as liabilities rather than assets and to get rid of as many of them as quickly as possible. They have also recommended slashing energy-efficiency programs that they see as nothing more than a customer-service frill. Often these two recommendations go hand-in-hand. The consultants have been aided and abetted in this disabling strategy by some Wall Street analysts who believe that the *sine qua non* measurement of economic competitiveness in the electric services industry is the price of something no one has ever seen, heard, or tasted—a kilowatt hour.

Unfortunately, all too many utility executives have been confused by the torrent of loose rhetoric and wishful thinking about retail competition, direct access, and consumer choice, leaving them frozen in the headlights. Some have huddled together in the middle of the road, believing that a herd is less likely to be run over by change, even if they're going the wrong way. Regardless of the riskiness of their response, their instinct is correct that something important is happening. Change is indeed coming to the electric services industry, and it will be even more wrenching and far-reaching than advocates of multiple vendors of retailed electrons are imagining.

The convergence of several technological, economic, and social trends will make the traditional utility model (i.e., large central power stations linked together by long, high-voltage transmission lines) obsolete sooner than almost anyone could have thought possible as recently as two years ago. The electric services industry will probably be configured more along the lines Thomas Edison envisioned a century ago than the systems built by his rivals George Westinghouse and Samuel Insull. Rather than hierarchical mega-monopolies commanding a brittle copper and aluminum web hooked to resource-intensive nuclear and coal plants, there will be resilient networks connecting manifold, diverse, and decentralized power plants. Many of these plants will be renewable, buffered by elegant small-scale energy storage systems; all will be managed by the distributed intelligence of local adaptive controls; and all will support customer devices that far more efficiently transform electricity into hot showers, cold beer, and other desired services.

The technologies driving these changes include fuel cells, photovoltaics, carbon-fiber flywheels, ultracapacitors, smart energy management systems, two-way communication technologies, and equipment that converts far less electricity into far more and better services that people want, such as comfort, security, safety, and fun.

These technological changes will be augmented by an opening up of the electric services business to a more diverse group of entrepreneurial actors. Their entry into the market will be driven by an increasing recognition that the electric services industry needs to be focused more through the windshield of economics rather than the rearview mirror of accounting. The revelation of the true costs of delivering electricity differentiated by time and location will encourage the introduction of many new technologies on both sides of the existing electric meters, and it may even lead to replacement of these meters with devices that give customers more control over how much they spend for electricity.

These technological and economic trends will be accelerated by citizen activists who are seeking ways to revitalize neighborhoods, stop environmental degradation, reduce social anomie, and create local economic opportunities. New energy technologies can give citizens the tools to improve local air quality, reduce urban sprawl, create jobs, and reverse the disinvestment that has occurred in many older neighborhoods and cities.

Ed Smeloff and Peter Asmus in *Reinventing Electric Utilities: Competition, Citizen Action, and Clean Power* bring a much needed perspective to the raging debate about the restructuring of the electric services industry.

More importantly, they have pointed out how some utilities and communities have embraced the positive trends that are already decentralizing the power system. The remarkable lessons learned at the Sacramento Municipal Utility District need to be widely shared with others interested in helping to shape the future of the electric services industry.

The authors' analysis of the California debate over restructuring is also instructive. There is much mythology about the 1994 California tsunami that was to "deregulate" electricity and create untold consumer benefits. Smeloff and Asmus probe behind the ballyhoo. They combine a keen political understanding of the interest groups involved in the California debate with their insiders' view of how electric utilities really work. They also clearly point out the key issues that still need to be resolved in the transition to a more competitive and community-oriented industry.

The debate about how to manage the transition to a decentralized power system will undoubtedly continue for years to come. Some will resist the changes and others will passively watch as more and more embrace the future. The changes will not occur overnight, but they will come faster than many expect. The current system will coexist with the new system for a while in order to ensure reliability of electric service. That may be long enough for a graceful transition. However, that is not inevitable. It will only occur if enough utilities and other actors in the electric services industry comprehend and grasp the extraordinary new opportunities now moving from the lab bench into the marketplace. *Reinventing Electric Utilities* points out how some of the utility leaders of today and tomorrow are already well on their way.

Amory Lovins
Director of Research
Rocky Mountain Institute
Old Snowmass, Colorado

Acknowledgments

Through its Energy Outreach Program, the Safe Energy Communication Council (SECC) has conducted more than 20 state media and speaking tours with Ed Smeloff to spread the message about Sacramento's success and the need for change in America's utility and energy systems. SECC was assisted in this effort by generous support from The Joyce Mertz-Gilmore Foundation. *Reinventing Electric Utilities* was completed with SECC's assistance.

The Energy Foundation, working through the Center for Energy Efficiency and Renewable Technologies, underwrote much of the book research conducted by Peter Asmus, particularly those sections of the book devoted to regional case studies of reform.

Also of key assistance were the following individuals and organizations who provided comments and/or important research in the development of the book: Scott Denman, Christina Nichols, V. John White, Eric Heitz, Ralph Cavanagh, Armond Cohen, David Moskowitz, S. David Freeman, K.C. Golden, Tom "Smitty" Smith, Pat Wood, Jim Caldwell, Martha Ann Blackman, *The Sacramento Bee*, California State University at Sacramento, Renewable Northwest Project, and the California Energy Commission.

Introduction

"Seek simplicity and distrust it."
—Alfred North Whitehead

These are unsettling times for electric utilities and the public they serve. In early 1996, nearly 40 states were considering fundamental changes to laws and regulations that gave rise to electric utility monopolies. Several state public utility commissions have released detailed proposals for ending the traditional relationship between state governments and electric utilities that requires regulated monopolies to provide universal electric service in return for predictable profits.

The new reforms are driven by the belief that more competition will make the production and delivery of electricity more efficient. Some states propose breaking up existing electric utility monopolies and creating new entities for the generation, transmission, and distribution of electricity. Congress is also considering the revision of several important federal laws to promote competition in the electric power industry.

Increased economic competition has affected many American industries over the past 20 years. Technological innovation, the increasing competitiveness of international markets, and ideological and political pressure are all behind the current drive for more reliance on markets and less on government regulation. Advocates for the restructuring of the electric service industry often point to the natural gas and telecommunications industries as

1

models of successful regulatory reform. Changes in these industries do offer some useful lessons, but there are also some important differences that need to be considered in the debate over restructuring electric utilities.

Deregulation of the natural gas industry has resulted in substantial new investments in gas exploration and recovery. Vast new gas reservoirs have been developed in the past decade. The price of natural gas has been lowered and remains stable. However, unlike electric utilities, the natural gas industry did not have to be restructured to introduce economic competition. Likewise, jurisdictional overlap between the federal and state governments was not a significant issue in the natural gas industry, while it is emerging as a major area for battle in electricity. Deregulation of the natural gas industry was accomplished by an act of Congress and a ruling of the Federal Energy Regulatory Commission (FERC), whereas electricity reforms will necessarily involve state regulatory commissions and legislatures.

In the telecommunications industry, deregulation has resulted in the introduction of a variety of new products and services and has lowered the cost of long-distance telephone service. The breakup of the Bell system was triggered by technological innovation. The introduction of microwave communication broke AT&T's monopoly on long-distance telephone service and opened the door for competing companies to enter the market. New growth in services brought on by faxes, answering machines, modems, and the Internet have been a bonanza for the telecommunications industry. This growth has allowed for the rapid introduction of new technologies. There is no corresponding growth seen for electricity consumption. In fact, many states and the federal government have encouraged reduction, not an increase, in the consumption of electricity.

Two critical characteristics of the electric power industry make it different from other industries that have been deregulated. First is the industry's historical capital intensity. Investments in a single nuclear reactor have reached $5 billion and can represent 40 percent or more of the total assets of the electric utility owner. The federal Department of Energy estimates that nuclear reactors represent about 47 percent of total electricity-generating assets in the United States, yet they account for only 22 percent of electricity generation in the country. No other industry has such a concentration of investments in facilities that contribute so little value for its customers. Dealing with these investments is the major challenge in creating a competitive electricity market.

These nuclear power plants also share a characteristic with other types of power plants that makes electric utilities unique. No other industry leaves such large and deep environmental footprints. The buildup of radioactive waste from nuclear reactors is just one impact. There are several others. Large fossil fuel plants produce 72 percent of the nation's sulfur dioxide, which is the principal cause of acid rain and a major source of smog. The burning of fossil fuels to generate electricity accounts for 36 percent of U.S. emissions of carbon dioxide, which contributes to global climate change. The world's leading atmospheric scientists warned in early 1996 that failure to curtail reliance on fossil fuels jeopardizes any credible response by the industrialized world to deal with this, the most threatening of all environmental challenges.

Besides these large-scale issues are the more visible regional environmental problems, such as mercury contamination of lakes and rivers and severe land erosion from strip mining of coal. The restructuring of the electric power industry could seriously exacerbate environmental problems, global and local alike. On the other hand, it could be part of a strategy that lessens the effects of electricity production on the environment through the introduction of cleaner power sources.

This book argues that the huge capital investments and corresponding ecological burdens in nuclear and coal plants need to be addressed head-on to establish a more sustainable long-term structure for the delivery of electric services throughout the United States. This means that it will be necessary to phase out uneconomic power plants as part of a transition to a more market-oriented electricity system.

And while the early retirement of noncompetitive power plants can be part of a strategy to lower electric bills, it is not sufficient for protecting the environment. This book argues that is is also necessary to devise policies to accelerate the introduction of renewable and other clean technologies into the mix of resources used to produce electricity. While competitive markets will be the preferred way of allocating resources in a reinvented electric services industry, there will continue to be a need to monitor the environmental impacts of market decisions and to mitigate them as necessary.

A review of the evolution of the electric service industry from a decentralized system to one of monopoly is the subject Chapter 1. A number of case studies follow, including a look at the Sacramento Municipal Utility District (SMUD). These case studies show how communities have organized to advocate for more sustainable energy policies to meet local and regional needs. Strategies in which citizens have a larger and more direct voice are highlighted, as are programs that deliver environmental benefits.

SMUD is the most significant case study and therefore will be described in the most detail. In 1989, SMUD's owners, the voters of Sacramento, California, closed the Rancho Seco nuclear power plant. In the seven years since the plant's closure, SMUD has maintained stable electric rates, improved its financial standing, and implemented a broad array of energy efficiency and renewable resource programs. The SMUD story demonstrates that competition can, if wedded to public processes that incorporate the values of local consumers, deliver a cleaner power system. The innovative sustainable energy strategies launched at SMUD, including the nation's most advanced solar "green pricing" program, can be replicated in other communities.

We recognize that there is not a single blueprint for reinventing electric utilities. However, there are common themes in the national debate about how to reinvent electric utilities. The first detailed proposal for restructuring the electric services industry was made in California. This book scrutinizes California's restructuring proposals, since they touch on a broad range of policy issues now confronting regulators at all levels of government across the country. It also examines the debate in New England, focusing on the collaborative effort of multiple stakeholders to set common restructuring goals. The debates in these regions of the country reveal some key principles related to restructuring of the electric services industry. However, specific solutions will vary depending on an area's resource needs, cultural values, and utility structure.

To augment the SMUD, California, and New England case studies, the book highlights the efforts of environmental and consumer organizations in other parts of the country to stop or phase out unproductive and environmentally damaging sources of power and to promote the use of new, cleaner energy technologies. These cases include the work of citizen groups in the Tennessee Valley, the Pacific Northwest, and the states of Texas and Minnesota.

The electric services industry in the United States is quite decentralized, especially in comparison with other large industries like petroleum refining, automobile manufacturing, and steel production. A diversity of decision-making forums at various levels of government and in the new marketplace can create opportunities for local communities to shape energy policies that can have important consequences for the economy and the environment. New coalitions of stakeholders, encompassing forward-looking public and private utilities, environmentalists and industries, and large and small consumers, may find that they can help reinvent electric utilities.

The future configuration of the industry is hard, if not impossible, to predict. There are forces pushing for consolidation of economic power in the electric utility industry. Certainly, the wave of utility mergers that began in 1994 is an indication that some utility executives believe that larger organizations—perhaps oligopolies—will be better able to weather future changes in the industry's structure. On the other hand, new technologies like combustion turbines, fuel cells, and photovoltaics may encourage further decentralization of electricity production.

Electric utilities will not be reinvented solely by technological leaps or by the magic of the market. Although the influence of these dynamic forces needs to be understood, the future ultimately will be determined by the ideas, hopes, and vision of people. This book is written for those who are engaged in the debate about the future of the electric services industry or who want to be. The debate is too important to be ignored because of complexity or to be left to utility experts or special interests. The success that Sacramento citizens had in helping to reinvent their electric utility is one example of what can be done. Others will surely follow.

Chapter 1

The Growth of Electric Monopolies

"Then there is electricity, the demon, the angel, the almighty
physical power, the all pervading intelligence!"
        —Nathaniel Hawthorne, *The House of Seven Gables*

For most Americans it is hard to imagine life without electricity. Yet in the
span of one lifetime, technologies that produce, deliver, and use electricity
have spread to nearly every community across this country. This bundle of
technologies has changed the culture and the economy of the United States
in ways that would have been unimaginable to Thomas Edison.

Personifying this remarkable transformation is the life of blues musician
B. B. King. In a speech to the National Press Club, broadcast on C-SPAN
in early 1996, the 70-year-old King said he had lived without electricity until
he was 18 years old. During those years, without the distractions of radio or
TV, he learned to play the guitar. Fifty-two years later King has released a
CD-ROM that with sound and images documents his life and the history of
the blues in America.

It is not hard, of course, to identify the many ways that electricity has
changed modern life for the better. It lights up the night, creates comfort in
inhospitable climates, and frees men and women from much drudgery. It
has made it possible for much of humanity to have immediate access to
enormous quantities of information. It is not an exaggeration to say that
electricity has even changed our very conception of space and time.

Yet electric power is not without costs. It has transformed the physical environment we inhabit. Dealing with the by-products of electricity challenges our imagination and creativity. The waste created from nuclear fission will remain hazardous for periods of time far longer than the history of civilization. The massive combustion of fossil fuels has already altered the chemical composition of the earth's atmosphere, changing global weather patterns in unpredictable and perhaps frightful ways. The correlation between rising carbon emissions and increases in average global temperature raises alarm (see Figure 1-1). While humans may be able to adapt to these changes, the effects on many species of plants and animals could be irreversible.

At a regional level, power plants burning coal in the Midwest affect the acidity of lakes in New England. Huge bodies of water, like the Great Lakes, are threatened by mercury contamination. In our urban metropoli, the health of millions is harmed by the smog created by the combustion of fossil fuels. These environmental issues loom large in the debate about how to continue wiring the world for electricity.

The per capita consumption of electricity in many developing countries is less than one-tenth of what it is in the United States. If our current patterns

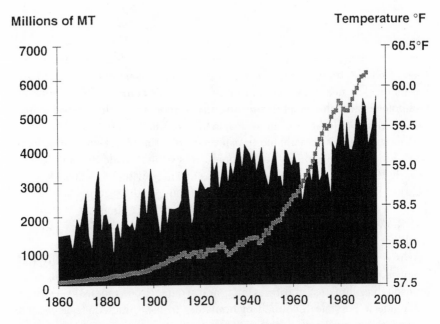

Figure 1-1: One hundred and sixty-two nations have committed in the Rio Treaty to help stabilize world carbon dioxide emissions, the primary gas associated with industrial activity and global climate change.

of electricity production and consumption are extended to these places, children born today could live in a world with five times as many large power plants. Two billion people, today, live in unelectrified villages. It is morally untenable, and politically impossible, to deny people living in the developing world the benefits of electricity. The challenge for humankind is to create institutions and policies at the local, national, and international level that can stimulate models for sustainable electrification. The wisdom needed to create these models will come, in part, from understanding how the electric service industry evolved in a democratic society like the United States and how it is being transformed by the forces of competition and citizen action.

Electricity Comes to America

The electrification of the United States did not occur overnight. It has a history. There were the early inventors who envisioned uses for electricity; entrepreneurs who struggled to create markets for electricity; engineers who patched together the webs of power lines and sought out improvements in power plants; and the large corporations that financed and organized the growth of the industry. In the six decades it took to wire America, the country was transfigured. People moved to cities, assembled huge factories, organized labor unions, built trolleys and subways, developed systems of instantaneous communication, and created new forms of mass entertainment, including radio, television, and computer networks.

A new institution was also born in these times, the electric utility. At first, electric service was a neighborhood affair. Hotels, department stores, and theaters were among the first establishments to be electrified. Often, power plants were located at the site where the electricity was used. Early electric grids were haphazard and outages were common.

During the heyday of Thomas Edison, the demand for electricity was so little and sporadic that it was difficult for power plant owners to make a profit. Edison and others spent much of their time coming up with new ways for customers to consume electricity. Growth in electric power consumption didn't take off until power lines were hooked up to trolleys. By the 1890s there was fierce competition among hundreds of companies to build trolley lines and electric power plants all across America. In just a few years electric trolleys were operating in 850 cities. Each city, of course, has it own story of electrification.

Electricity was first delivered in Sacramento, California, in 1895. On July 13th, electrons vibrated over a 22-mile power line from a hydroelectric power plant stationed by the town of Folsom on the American River to downtown Sacramento. The *Journal of Electricity* proclaimed Sacramento to be "the first American city to demonstrate . . . long-distance transmission at high voltage." Thirty thousand people came to Sacramento on September 9th—California Admission Day—to see the rotunda of the State Capitol building emblazoned by electric lamps.

The business of supplying electricity to cities like Sacramento would become quite profitable. Competition for those earnings spawned strife. At the turn of the century, conflicts between electric companies and with city governments were the rule of the day. Some local political leaders wanted to create municipal power systems. In some cities, fights were waged by private companies to obtain exclusive service franchises. Often, bribes were paid, or other favors granted, to local officials to secure the franchise. Politics was part and parcel of the electric power business from day one.

The specter of public ownership of electric systems and the oftentimes messy rivalries between private companies were seen as baleful developments by some leaders in the electricity business. They threatened the budding profitability of an industry that needed stability. The advent of alternating current made it possible to transmit power over long distances, enabling power plant owners to greatly expand the range of services offered.

Companies like Edison General Electric in New York, Chicago Edison, and Southern California Edison saw opportunities to build big markets for electricity by coupling large power plants to high-voltage power lines. This combination promised low-cost power to consumers and rich returns to utility shareholders.

State Regulation of Utilities

State regulation of electric utilities was the brainchild of Samuel Insull, a protégé of Thomas Edison and later the owner of Chicago Edison, now known as Commonwealth Edison. At the turn of the century, Insull told other utility executives that competition was not in the best interest of the electric business. He observed that competing power lines and power plants increased the costs of delivering electricity. The business of providing electricity was, in his judgment, a natural monopoly. The question at the time was who would monopolize the sale of electricity: local governments or private companies?

Insull told his colleagues that a reasonable alternative to public ownership was to grant state governments the power to regulate their businesses to allow for predictable profits. For many businessmen the idea of government regulation was anathema. But it was a view that soon prevailed. In 1907, the National Civic Federation, a group of prominent businessmen that included

Insull and the well-known banker J.P. Morgan, released a study of the electric power industry. It called for the establishment of a system of electric monopolies that would be regulated by state public utility commissions. The idea of electric utilities as natural monopolies requiring state regulation soon dominated the way electricity was produced and distributed in the United States. State regulation created a stable legal and economic structure that assured the rapid growth of vertically integrated utilities that ran everything from power plants to high-voltage transmission to distribution wires in their exclusive service areas.

For almost 90 years this model of a regulated electric monopoly flourished in the United States. During its reign hundreds of large power plants were built and tied into grids of high-voltage transmission lines that eventually stretched across America's vast landscape. Most of the electricity sold today in the United States still comes from large coal, nuclear, and natural gas power plants located outside of urban areas and delivered to population centers over high-voltage transmission lines. Only recently has that structure begun to change.

Public Ownership

Since electricity was introduced in the United States at the turn of the 20th century, the only serious challenge to control of the electric power infrastructure by private monopolies occurred during the administration of Franklin Delano Roosevelt. As governor of New York, Roosevelt was outraged by the high electric rates charged by the state's electric utilities.

An alternative to privately controlled electric utilities was needed, he thought. Roosevelt decided that the hydroelectric potential of the St. Lawrence River should be developed by a public agency. To accomplish this goal he created the Power Authority of the State of New York. This institution was to serve as a model for Roosevelt once he became president.

During Roosevelt's "New Deal" in the 1930s, the federal government launched a vast electrification program that included the creation of huge federal power agencies. These agencies were created to develop the most accessible renewable source of energy at the time—the nation's rivers. Large federal dams were also built by the U.S. Department of the Interior in the Southwest and in California. City and county utilities and other public power agencies were given the right to purchase electricity from these federal institutions before it could be marketed to private utilities.

For Roosevelt, public power systems were to be the "yardsticks" that could be used to measure whether private utilities were price gouging. He observed that public power could be used "to prevent extortion against the public and encourage the wider use of electricity." Campaigning in 1932, Roosevelt argued that if the citizenry was not satisfied with the service of a private utility, it had the right to set up its own governmentally owned and operated utility. Many communities did just that. Roosevelt's view that public power could be a tool for economic development was warmly embraced in the Pacific Northwest and the Tennessee Valley.

The development of a renewable source of electricity from the nation's waterways was also spurred on by the Federal Power Act, which had been passed by Congress in 1920. This law gave municipal utilities preferential treatment for hydroelectric development. The law encouraged community leaders in Sacramento to form a municipal utility for the development of water and power for California's capital city.

Sacramento was originally a gold rush town. It grew at the confluence of two powerful rivers, the Sacramento and the American. The Sacramento was one of the West's largest navigable rivers, meandering from north to south through California's Central Valley. The three forks of the American rush down from the snow-capped Sierra Nevada mountains, depositing gold-laden sediment along river banks before merging with the Sacramento River. Summers in the valley could be miserable, especially when the American was low and the Sacramento was filled with the effluence of upstream towns. The people of Sacramento often complained of putrid drinking water. It was a desire for pure mountain water that convinced community leaders to form a municipal utility district. They hoped that the generation and sale of electricity would provide the revenue to pay for the dams needed to store water in between rainy seasons.

Sacramento's leaders wanted to take over the electric distribution systems of Pacific Gas & Electric (PG&E) and the Great Western Power Company and to construct a series of hydroelectric plants on the American River. But their hopes were blocked by litigation with PG&E, which, by the mid-1930s, had become the monopoly supplier of electricity in Sacramento. PG&E opposed every step Sacramentans took to form a municipal utility and to develop water and power supplies. After 23 years of legal wrangling, Sacramento citizens finally prevailed on New Year's Eve 1946, when the Sacramento Municipal Utility District (SMUD) took over the operation of Sacramento's electric distribution system.

The 1950s and 1960s were a time of building for utilities throughout the country. The American River offered an abundant supply of renewable hydroelectric energy for Sacramento. It took SMUD 13 years to build the Upper American River Project, a 40-mile stairway of power consisting of 11 major dams, 6 powerhouses, and 24 miles of tunnels. At the time these hydroelectric projects were built they generated electricity at a cost roughly equivalent to that produced in PG&E's large oil-fired power plants. But the dividends from the renewable energy project have proved to be long-lived. Today, these hydroelectric power plants provide the lowest cost electricity in the SMUD system.

Public power agencies such as SMUD developed many of the hydroelectric power plants now operating in this country. However, the potential for hydropower was largely tapped out by the mid-1960s. To obtain additional power most publicly owned utilities, like their investor-owned cousins, decided to build large nuclear and coal power plants.

The End of Stability

In the 1970s, after more than 50 years of stable growth, several disturbing trends developed that began to erode the way electric utilities, public and private, carried out their responsibilities. The utility strategy of building bigger and bigger power plants had made economic sense as long as the next plant was cheaper than the last one and as long as rapid growth in demand for electricity continued. By the 1970s these two conditions were beginning to change. Bigger was no longer better, and consumers of electricity were beginning to find ways to get more from less.

In the 1930s, the most efficient power plant produced about 50 megawatts of power, much more than any single consumer of power could use. It therefore made sense to serve many customers from one central power supplier. Engineers were finding that further economies could be achieved by building larger scale power plants. For the next 40 years regular improvements were made in steam turbines, which were the workhorse of power generation technology. More efficient, larger power plants steadily replaced smaller, less efficient facilities. It made sense for utilities to encourage customers to use more power in order to finance even bigger and cheaper power plants. It was a time of extraordinary stability. Rates were declining in real terms and profits were increasing.

By the mid-1970s, the lowest cost power plant was 20 times larger than one built in the 1930s. These gigantic dynamos could produce 1,000 megawatts of electricity, enough power to serve a good-sized city. Despite years of steady improvements, large thermal power plants are not very efficient. A large coal plant converts about one-third of the energy contained in the coal into electricity. The rest of the energy is lost as waste heat.

Engineers have worked hard to improve the efficiency of large-scale power plants that boil water to make highly pressurized steam that drives a turbine that is connected to an electric generator. The improvements over the last 20 years have been, nonetheless, marginal. During that time, the efficiency of all thermal power plants in the United States grew from 29 percent to just over 30 percent. That means that 70 percent of the energy contained in fuel is wasted.

As the efficiency of these plants leveled off, so did their costs. Utility executives and energy policymakers began to look for alternative ways of lowering the cost of electricity. Many believed they would be able to further drive down the cost of electricity by using nuclear fuel instead of coal to boil water to power the steam turbine. Among them was Paul Shaad, SMUD's general manager in the 1960s. He had been interested in nuclear power since the end of World War II and received a security clearance from the federal government to obtain classified information about the development of this new and promising source of electricity.

In 1963, General Electric (GE) made a bold move to convince utilities that nuclear power had arrived as a cost-competitive alternative to fossil fuel. The company won a bid to build a nuclear plant for the Jersey Central Power and Light Company by offering a fixed-price contract of $66 million. At that price nuclear power was more economic than coal. President Lyndon Johnson called the contract an "economic breakthrough" for nuclear electricity. SMUD's Paul Schaad was also impressed. According to Shaad, the contract "showed that nuclear energy was competitive with fossil fuel in the East, and more than competitive with gas and oil here in California." The cost of electricity from nuclear power was estimated to be less than half a cent per kilowatt hour.

Soon after the sale of the New Jersey reactor, Shaad attended a symposium sponsored by the Atomic Energy Commission, the federal body originally set up to regulate nuclear materials (later it was to become the Nuclear Regulatory Commission). Representatives of GE, Westinghouse, and Babcock and Wilcox told utility executives that they were open for business and ready to accept orders for a low-cost technology that would meet America's power needs far into the future. In 1964, SMUD ordered what would turn out to be the 50th nuclear power plant in the United States. A site was found in a sparsely settled area of rolling hills, 25 miles southeast of Sacramento. In 1966, SMUD purchased a tract of land called Rancho Seco, named after the title of the original Spanish land grant. SMUD contracted with the Bechtel Corporation to design and supervise the construction of the plant. Construction began in 1969 and was completed in 1974. The original cost estimate for the plant was $180 million. The price would be double that by the time Rancho Seco was completed in 1974.

As nuclear technology was being introduced in Sacramento and at dozens of other utilities across the country, several economic and social trends began to shake up the power industry. In the late 1960s inflation emerged as an important problem for the U.S. economy, brought on by spending on the Vietnam War and Great Society programs. Inflation exceeded 5 percent in 1971, almost double what it had been in the 1950s and early 1960s. By 1980 it would reach 18 percent. The cost of borrowing money rose with high inflation. For electric utilities, which are among the most capital intensive of all businesses, high interest rates put pressure on electricity costs.

Then in 1973 came the Arab oil embargo and the emergence of OPEC as a cartel that would manipulate petroleum supplies to control prices. Crude oil jumped from about $3 per barrel to almost $12. After the Iranian revolution in 1979, the price went up to over $30 per barrel. Prices for natural gas followed this trend, quadrupling in less than a decade.

For electric utilities using large quantities of oil and natural gas to produce electricity, these events meant trouble. In the short term, utilities had no choice but to raise electric rates and customers had little choice but to buy this higher cost electricity. The common perception at the time was that oil prices would continue to rise rapidly. OPEC's stated policy of supporting price increases of at least 10 percent per year fed that concern. Some observers assumed the price of oil would reach $100 per barrel by the end of the century. The fear of very expensive imported oil caused energy policymakers to demand that electric utilities turn to coal and nuclear as the preferred way of meeting the nation's rapidly growing need for electricity.

The 1970s also marked the beginning of the environmental movement as a significant force in U.S. politics. The first Earth Day in 1970 tapped into strong public sentiment about a deteriorating environment. The National Environmental Policy Act was passed in short order by Congress and signed into law by President Richard Nixon. By administrative fiat Nixon set up the Environmental Protection Agency. In California an even stronger environmental law, the California Environmental Quality Act, was signed by then Governor Ronald Reagan. These laws mandated comprehensive environmental assessments of any infrastructure projects that required government approval.

These and other laws, such as the Clean Air Act, increased the time and costs of developing new power stations. The Clean Air Act required owners of power plants to install air pollution controls to meet new federal standards intended to improve public health. The effect of this law on the cost of electricity was most notable for coal-fired power plants. At one large coal-fired facility built in Arizona in the 1960s, the cost of air pollution controls added in the 1970s equaled the original power plant cost.

All of these factors led to large rate increases. Utilities' marginal costs—the cost of producing additional power—became higher than average costs for the first time. This reality meant that growth in demand for electricity would trigger higher prices. However, most utilities in the 1970s had rate structures that encouraged the consumption of more electricity. The more you consumed the lower the price. The combination of rising marginal production costs and declining prices for consuming power was quickly eroding the financial well-being of many utilities.

A Time of Denial

Many electric utilities were paralyzed by these developments. They refused to see that the cross-over in marginal and average costs had caused their businesses to change dramatically. Their response was to appeal to state regulators for more rate increases. Despite rapidly rising rates, many utilities still assumed that the demand for electricity would grow at the pace it did during the 1960s. Most continued to order large power plants to meet the expected growth in customer demand. Many turned to nuclear power in the expectation that this technology would be a low-cost source of electricity, since uranium was widely available and cheap.

The reliability of the electric system was a top consideration for most utility executives. They feared a repeat of the 1965 Northeast blackout. It was not unusual at the time for utilities to forecast growth in demand for electricity at 7 to 8 percent per year. That would mean a doubling of the number of power plants in about a decade. In 1970, Southern California Edison, the big utility serving the greater Los Angeles area, planned to build seven nuclear plants and a coal plant by the 1980s.

As utilities tried to implement these resource plans, rates went through the ceiling and triggered a backlash among consumers. Electric rates in California, for example, would more than double during the 1970s. Grass roots organizations like Fair Share, Acorn, Citizens Action, and the Public Interest Research Group (PIRG) set up chapters across the country to fight rate increases. Under consumer pressure many states created departments within their attorney generals' office or at other state agencies to represent consumer interests during rate hearings.

This surge of consumer pressure forced many public utility commissions to cut back on utility requests for rate increases. *Lifeline rates,* which provide a lower rate for a first block of electricity and a higher rate once that amount is exceeded, became an important political issue in many states. The idea was to protect customers who use electricity to meet basic needs and to provide a price signal to encourage consumers to conserve.

Rate increases and policies encouraging energy conservation began to curb the rate of growth in demand. Instead of a 7 percent per year increase in consumption, demand growth quickly dropped to about 2 percent per year. With this dramatic decline in growth, many power plants that were being built were no longer needed. However, some utility planners saw this decline as only a temporary phenomenon and predicted that economic growth would by necessity require a rapid growth in the amount of electricity produced.

At the same time, technologies that used electricity more efficiently to produce lighting, refrigeration, heating, cooling, and ventilation were beginning to enter the marketplace. It was possible to build homes that could be heated for less than a dollar a day. Architects were also able to design commercial buildings that used half the energy they did a decade earlier. The more widespread application of these technologies and skills was limited, however, by imperfections in the regulated market for electric services.

According to conventional economic theory, consumers should invest in efficient levels of energy conservation on their own, motivated by self-interest. After comparing alternative investment choices, they should choose to pursue all energy efficiency investments that provide better rates of return. This model of consumer behavior assumes that market barriers are negligible or affect only a small number of people. However, research by the Electric Power Research Institute (EPRI) has shown that institutional barriers, inconvenience, risk perceptions, and limited access to capital and information had severely limited investments in energy efficiency measures.

These barriers to the more rapid introduction and widespread use of more efficient technologies are caused, in part, by the monopoly structure of the electric power industry. State-regulated electric utility monopolies are guaranteed an opportunity to earn a reasonable rate of return on prudent investments. Real rates of return on equity are in the neighborhood of 5 to 7 percent, and about 2 to 4 percent for debt-financed investments. Since those profits are virtually guaranteed once an investment has been deemed prudent, electric utilities can afford to amortize their investments in power plants over a relatively long period of time. Since a power plant could be used for 30 years or even longer, it is not unusual to wait 15 to 20 years before a power plant is fully depreciated.

For most businesses and energy consumers, 15 to 20 years is a very long "payback" period. Surveys of consumers have shown that they are unlikely to make investments in end-use efficiency unless the payback time is short—six months to three years. The payback on an efficient lighting system for an office building might be five years. That is, the amount of money saved by buying less electricity for lighting is enough to pay off the cost of the new lighting system in five years. If it takes 15 years to pay off an investment in a power plant to produce a certain quantity of power but only 5 years to pay off an investment that can save that same amount of electricity, it makes sense from a societal point of view to invest in the energy-saving technology.

Because of the different perceptions of risk between regulated electric utilities and businesses operating in competitive markets, that sensible investment is often not made. This market imperfection has been called the payback gap. This gap resulted in tens of billions of dollars invested in power plants instead of in lower cost energy efficiency measures.

Advocates of energy efficiency began to argue that utilities could correct this payback gap by providing customers with information and incentives to stimulate investments to reduce demand. And it could become a profitable activity if the link between the sale of kilowatt hours of electricity and utility revenues were severed.

Competition Comes to the Electric Power Business

President Jimmy Carter entered the White House in 1977 with the goal of increasing America's energy independence. He called for a national campaign to increase domestic sources of energy and to conserve energy. He called it the moral equivalent of war. Although Carter was stymied by Congress in implementing some of his ideas, he succeeded in passing three key pieces of energy legislation.

The Power Plant and Industrial Fuel Use Act was the first law signed by Carter. It prohibited the use of oil or natural gas in new electric power plants. At the time, almost half of the oil America used for electric generation and transportation fuels was imported. It was also widely assumed at the time that natural gas was in short supply. There had been curtailments of deliveries of natural gas in the winters of 1976–1977 and 1977–1978. The intent of the law was to preserve natural gas for heating and other high value uses.

The second law to be enacted was the Natural Gas Policy Act, which deregulated the price of newly discovered natural gas. Price deregulation was intended to stimulate exploration for new sources of natural gas within U.S. borders.

The third law was the Public Utilities Regulatory Policy Act (PURPA). This law encouraged utilities and state regulators to review utility rate design and load management programs. Its most far-reaching part turned out to be an almost unnoticed provision of the law that dealt with nonutility producers of electricity. It required that utilities buy electricity from private companies when that was a lower cost alternative to building their own power plants. At the time it was thought that this law would stimulate the development of renewable technologies like solar, wind, biomass, and geothermal power.

However, PURPA also provided a major stimulus to a technology that had been around for some time but was not widely used by electric utilities.

That technology was the jet engine. Called a combustion turbine by electric utilities, it could be used to provide mechanical energy to drive an electric generator. Combustion turbines were to turn the economics of power generation upside down. But, first, they needed an opening to enter the electric power marketplace. The Power Plant and Industrial Fuel Use Act prohibited the use of natural gas in new power plants, yet this was the best fuel for combustion turbines. State regulatory incentives also encouraged utilities to construct capital intensive power plants like nuclear and coal plants. Although more expensive to build than a combustion turbine, they were thought to be less costly to operate.

Since utilities earn profits on their investments in power plants, but not on the fuels the plants use, it is not surprising that they chose to build expensive power plants with low fuel costs. The more money prudently invested in power plants the bigger the profits. However, PURPA changed the economics of power production. It allowed combustion turbines to be used in cogeneration power plants. Under this configuration, the exhaust heat coming out of a combustion turbine is recovered and used productively by a thermal host, which could be an oil refinery, a food processor, a brewery, a paper mill, a chemical manufacturer, a sewage treatment plant, a hospital, or even an office building (see Figure 1-2). The exhaust heat can also be captured to generate electricity using a steam turbine and electric generator.

Today, the most efficient commercial power plant is a 40-megawatt combustion turbine that was originally developed for a Boeing 747 airplane. When used in a cogeneration plant, 75 percent or more of the energy contained in the natural gas fuel can be captured for productive uses, a level of efficiency that dramatically lowers the cost of producing electricity. The economy of scale for the electric power industry had been turned on its head. Smaller was cheaper. Moreover, combustion turbines could be built quickly, they better matched load growth, and they caused less pollution. They were therefore less risky than a large nuclear or coal plant that required huge capital outlays.

In the 1970s the California Public Utilities Commission (CPUC), whose members had been appointed by then Governor Jerry Brown, began to question utility resource plans that called for the construction of dozens of new big power plants, including several nuclear reactors to be built along the Golden State's prized coastline. The immense amounts of capital required to build these power plants were causing consumers to rebel against the rate increases that were needed to pay off the debt. The CPUC began to require utilities to implement conservation programs before they could justify building new power plants. Utilities soon got the message. By 1981, California's

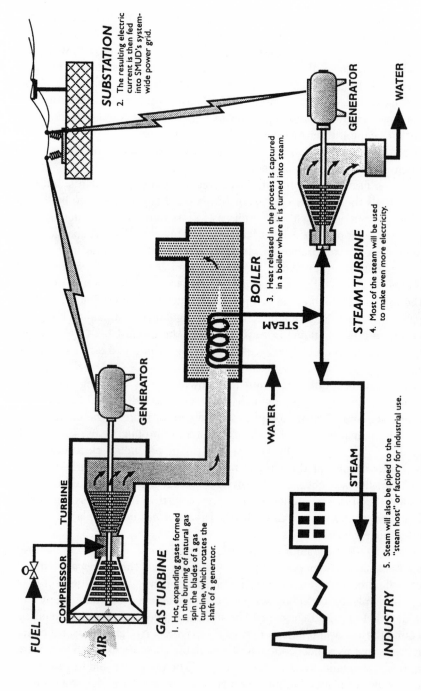

FUEL

AIR

COMPRESSOR **TURBINE**

GENERATOR

GAS TURBINE

1. Hot, expanding gases formed in the burning of natural gas spin the blades of a gas turbine, which rotates the shaft of a generator.

SUBSTATION

2. The resulting electric current is then fed into SMUD's system-wide power grid.

WATER

BOILER

STEAM

3. Heat released in the process is captured in a boiler where it is turned into steam.

GENERATOR

WATER

STEAM TURBINE

4. Most of the steam will be used to make even more electricity.

STEAM

INDUSTRY

5. Steam will also be piped to the "steam host" or factory for industrial use.

Figure 1-2: How cogeneration works. *Source:* Sacramento Municipal Utility District.

investor-owned electric utilities had become national leaders in energy conservation, spending about $50 million a year on programs for saving electricity.

At this same time, California's utilities were also clamoring for protection from the risk of rapid changes in the price of oil and natural gas. The risks California utilities faced from these price changes were heightened by the large amount of hydroelectric generation they owned. In a drought year, the swing in costs to displace cheaper hydro power was very large, affecting utilities' cash flow and ability to borrow money. At the time, California's three investor-owned utilities were having serious financial difficulties. San Diego Gas and Electric was on the verge of bankruptcy, and PG&E and SCE were in the midst of major power plant construction programs. The CPUC was sympathetic to this problem and decided to allow the utilities to pass through to customers any changes in the cost of obtaining fuel to run their power plants.

Another risk faced by utilities was the fluctuation in the sale of electricity to consumers from year to year. In California, utilities and intervenors representing consumers would argue endlessly over utility forecasts of sales because they affected rates charged for electricity. Rates were set by dividing the cost of providing electric service by a forecast of sales. This exercise was performed for each customer class: residential, agricultural, and industrial. A high sales forecast resulted in lower rates, and vice versa. Utilities had an incentive to come in with low forecasts to make rates higher. If sales were greater than the forecasts they made, they received bigger profits.

Successful energy conservation programs lowered electricity sales and profits. Utilities, therefore, had an economic reason to resist pursuing conservation. Environmentalists, working with state regulators, began to realize that investments in energy efficiency programs needed to be as profitable as investments in new power plants if they were to be successful.

To address this dilemma, the CPUC adopted a policy called the Energy Revenue Adjustment Mechanism (ERAM) in 1982. ERAM ensured that utilities would collect the revenue projected in their sales forecast. If utilities collected more—due to increased sales—it was refunded to the ratepayers in the next year. If they collected less, a surcharge was added. With ERAM, utilities would not lose money when conservation programs were successful.

California also developed efficiency standards for appliances and buildings. Appliance standards for electrical equipment like refrigerators and air conditioners became *de facto* industry benchmarks since national manufacturers could not ignore the large California consumer market. The federal government later adopted many of the standards California pioneered. The California Energy Commission estimates that by 1999 utility conservation programs and the state's building and appliance efficiency standards will have conserved more energy in California than could be produced by 12 large nuclear power plants.

While California's actions prodded federal officials to institute national energy efficiency standards, it was PURPA that introduced competition into the generation of electricity. In 1982, the CPUC authorized standardized contracts to implement PURPA, which pegged payments for electricity to projections of oil prices. Long-term contracts were essential for independent power producers to attract financing. California's contracts proved to be very lucrative and jump-started the country's development of alternative sources of power.

California utilities, which initially fought PURPA, were placed on the defensive when the nuclear plants they were building faced significant delays and huge cost overruns. The most notorious was PG&E's Diablo Canyon nuclear facility, where blueprints for earthquake safety structures were read backwards. The two-reactor facility took over 18 years to build and cost more than $5 billion. While not attracting as much attention, setbacks also occurred during the construction of the San Onofre Nuclear Generating Station (SONGS) plants in Orange County and the Palo Verde facilities north of Phoenix, Arizona, that Southern California Edison had bought into. State energy policymakers were concerned about the possibility of shortages in energy supplies if these nuclear facilities continued to be delayed. To meet the potential shortfall, the CPUC ordered the state's investor owned utilities to buy power from independently owned power plants.

This policy helped establish an industry that is now building renewable power plants throughout the world. All told, about 11 percent of California's electricity is now supplied by a mix of non-hydro renewables, including the largest solar generating capacity (374 MW), wind capacity (1,812 MW), geothermal capacity (918 MW), and biomass capacity (1,031 MW) in the world.

Utilities in other parts of the country were slow to see the advantages of the smaller scale technologies that were being installed in California. Many continued to believe that coal and nuclear power were the only alternatives to meet future power needs. A prominent group of policy analysts at Harvard's Kennedy School of Government wrote an influential book about the crisis in the electric utility industry in 1985. Entitled *The Dimming of America,* it concluded that regulation of electric utilities in the United States had failed. It warned that if changes were not enacted quickly there would be blackouts across America in the 1990s. The thesis of the book was that the political nature of state regulatory bodies had suppressed the need for electric rate increases, which led to the undercapitalization of the industry. That meant that electric utilities would not be able to build needed power plants and that would inevitably lead to America's economic decline.

The book's prescription for avoiding the disastrous decline of America's economy was bitter but, in the view of the authors, necessary medicine. They called on the states to protect Public Utility Commissions from the political pressure brought to bear by consumers wanting to suppress rates. If that didn't work, the authors claimed the federal government would have to take over the role of overseeing electric utilities. They argued that this could be justified for national security and economic policy reasons.

At the time, the U.S. Department of Energy (DOE) embraced this analysis of the problem that electric utilities faced. They were concerned with the numerous delays and cancellations of new coal-fired and nuclear facilities. They argued that these plants were needed to reduce the cost of electricity by displacing high-cost fuels like oil and natural gas. Although DOE officials were convinced that the nation would soon face a crisis, they were unable to convince President Reagan that drastic action was needed.

Alternative views of how to meet the nation's future energy needs were beginning to gain credibility. Environmental groups like the Natural Resources Defense Council and the Conservation Law Foundation were beginning to map out a strategy that recognized how changes in technologies could slow the growth in demand for electricity and allow for it to be produced with less damage to the environment. These ideas awaited opportunities to be put into practice. In New England and the Pacific Northwest, the timing was right in the mid-1980s. Both regions had experience with troubled nuclear construction programs engendering large-scale public opposition. Citizens in these regions began demanding a say in decision making about utility resource plans, opening up the electric utility industry to more public scrutiny.

While these changes were occurring in the Northwest and in New England, SMUD was struggling with a troubled nuclear power plant, Rancho Seco. The plant was originally scheduled to go into operation in October of 1974. Several thousand people had gathered at the site to dedicate Sacramento's new source of electricity. President Gerald Ford had sent a telegram congratulating SMUD for helping to reduce the nation's imported oil by one million barrels of oil a day. What the crowd didn't know was that the plant's steam turbine had malfunctioned, forcing a shutdown before the ceremony began. Although the concealed problem didn't dampen the festivities on its dedication day, it turned out to be an omen of things to come.

Chapter 2

Pulling the Plug on
Nuclear Power in Sacramento

"We of the age of the machines, having delivered ourselves of
nocturnal enemies, now have a dislike of night itself."
—Henry Beston

In the early dawn of the day after Christmas, 1985, something went wrong
at the Rancho Seco nuclear reactor located on the outskirts of Sacramento.
At 47 seconds after 4:13 a.m., alarm bells sounded in the control room, the
nerve center of the gigantic nuclear power plant. At that moment, the operators had lost control of the 913-megawatt (MW) nuclear reactor that provided SMUD with most of its power. For the next 26 minutes, panic and
mayhem ensued.

To the astonishment of those on the plant's graveyard shift, the critical
system that controls the reactor suddenly went dead. The integrated control
system (ICS), which was supposed to keep the power plant in a precisely
tuned balance by controlling the pumps and valves that send cold water into
the reactor, had suddenly failed. The failure of this intricate computer system had never happened before at Rancho Seco, and the crew was not sure
what to do.

Automatically, valves and pumps set themselves at a halfway position. In
16 seconds the temperature in the reactor climbed from 582°F to 607°F.
With the increase in temperature, control rods automatically fell into the reactor core, stopping the nuclear chain reaction.

Once the plant stopped producing power, alarm systems began to blare. Unfortunately, the alarms signaled emergencies that had nothing to do with the actual problem. They included fire alarms, an earthquake alarm, and an alarm warning of rising temperatures in the spent fuel pool, where most of the highly radioactive material at Rancho Seco was stored. The control room was filled with a cacophony of noise and confusion.

In less than a minute after the reactor scram, the control room operators realized that cold water still flowing into the reactor core would send temperatures tumbling faster than safety regulations allowed. This was a potential catastrophe. If the reactor cooled too rapidly, the eight-inch-thick steel walls of the vessel containing the nuclear fuel could crack like a hot glass plunged into cold water. A breach of the reactor vessel would uncover the radioactive fuel, resulting in a core meltdown.

The control room operators knew the situation was precarious. But with the ICS out, they could not control the valves that allowed cold water to enter the reactor. Instead, they had to send workers scrambling throughout the plant—in some cases down three flights of stairs—to close valves by hand. But there, workers encountered new difficulties. The manual controls on the valves had not been properly maintained. Several of them stuck. Workers had to use wrenches to close some of the valves. One valve could not be closed.

An emergency heating and air conditioning system designed after the Three Mile Island accident in 1979 went on. But the system made so much noise that the operators were forced to shut it down just two minutes later. Six minutes after the ICS failed, the temperature in the reactor had dropped by 85°F. If this continued, the temperature would soon reach the point where the metal in the reactor walls would become brittle. If the reactor cracked, large quantities of radiation would be released into the containment building, the plant's last line of defense.

At 4:40 a.m., with the overcooling still not controlled, a senior reactor operator noticed that two switches in a recess above the control panel were set halfway to the off position, indicating that circuit breakers had cut power to the ICS. He flipped them on, 26 minutes after they had gone off. He then collapsed on the floor of the control room. An ambulance was called and he was taken to a local hospital where he was later released with a diagnosis of hyperventilation. The temperature of the water in Rancho Seco's core fell by 180°F in 24 minutes. NRC regulations require that a reactor vessel not be cooled at a rate exceeding 100°F per hour. That morning Rancho Seco cooled down almost four times that fast.

At 5:00 a.m., operators discovered that a reactor feedwater pump had been left running after water flowing into it had been cut off. The pump burned out and radioactive water spilled onto the floor of the auxiliary building that housed equipment connected to the reactor. Two workers entered the contaminated room and found three to four inches of water on the floor. They did not wear respirators, a violation of safety rules. Fortunately, they were not there long enough to be exposed to a dangerous level of radiation.

At 8:41 a.m., more than four hours after it started, the "unusual event," a term used by the NRC to describe the least serious of four categories of emergency situations, was declared over. But the ordeal for SMUD had just begun. The troubles at Rancho Seco would force the Sacramento municipal utility to confront issues no other utility in the country had faced until that time.

The Economics of Nuclear Power

The cause of the December 26, 1985, Rancho Seco accident was a crimped wire in a tiny electrical switching box. A short circuit had cut off power to the ICS, the computer that controlled the operation of the nuclear reactor. The problems at Rancho Seco, however, were much more extensive and severe and involved hardware, management, and training. However, the biggest challenge for SMUD would be the economics of nuclear power. At the time of the accident, SMUD did not believe that it had any other alternative but to repair the plant and get it running again. The cost of those repairs eventually brought the Sacramento utility to the brink of bankruptcy.

After investigating the accident, the NRC uncovered a multitude of problems at the plant relating to plant design, training, and communications. Still, SMUD management initially thought it could get the plant running in one or two months. However, the NRC disagreed and ordered the utility to conduct a complete review of every major system at the plant.

The financial institutions that had sold SMUD's tax-free bonds began to worry. They had been severely criticized for not exercising adequate oversight of other utilities that had made large investments in nuclear power. The two major credit rating agencies, Moody's and Standard and Poor's, announced that they were reviewing SMUD's AA bond rating after the NRC decision to delay the plant's restart. A lower bond rating would mean higher interest rates on the money SMUD borrows to finance construction.

A prolonged outage at the nuclear power plant would be a major blow to SMUD's ratepayers. Every day the plant was idle the utility had to purchase $262,000 worth of power. In addition, they were paying for the large work force that normally operated Rancho Seco plus a throng of contractors who were swarming over the plant trying to figure out what needed to be done for restart. In short order the costs of repairing the nuclear reactor began to strain SMUD's financial capabilities. SMUD's financing, like that of other public power agencies, comes from revenue bonds. In order to sell revenue bonds a utility must demonstrate that it has adequate income to pay them off. Income has to be sufficient to cover the interest and principal on the bonds plus a reserve margin. These fiscal fundamentals mean that large capital projects, like building or repairing a nuclear reactor, could create financial difficulties for utilities like SMUD.

At the time SMUD decided to build Rancho Seco it entered into a contract with Pacific Gas and Electric Company (PG&E), the nation's largest electric utility, which served the area surrounding Sacramento. That contract, called an integration agreement, allowed SMUD to finance the construction of Rancho Seco with tax-exempt revenue bonds. Both SMUD and PG&E foresaw long-term benefits from this arrangement. Surplus power from Rancho Seco would be sold to PG&E, and the operation of the two utilities' bulk power systems would be integrated into one system. PG&E would supply back-up power when Rancho Seco was closed for repairs or refueling, and SMUD would give PG&E the exclusive right to buy the low-cost surplus power from the plant that both parties assumed would be available in large quantities. The sharing of power reserves and the integration of a large geographical region, essentially all of Northern California, into one network was seen as a sound arrangement. As the cost of large power projects increased, the power industry saw this type of partnership between utilities as necessary to meet the growing needs for electricity.

The integration agreement served SMUD and PG&E well for the first seven years that it was in effect. A large quantity of the electricity that PG&E produced came from oil-fired facilities that had been built about 20 years earlier. With the increases in the cost of oil brought about by the 1973 and 1979 crises in the Middle East, it was becoming expensive for PG&E to operate these plants. The surplus power from Rancho Seco was, at the time, less costly than burning oil. Also, PG&E was concerned about power shortages in Northern California. The San Francisco–based utility was struggling to complete the two Diablo Canyon nuclear reactors that had been ordered at the same time as Rancho Seco. Delays were causing alarm that in the near future there might not be enough sources of power to meet the growing energy demands of Northern California. In the late 1970s and early 1980s the electricity from Rancho Seco was seen as essential for maintaining a reliable power grid for an area inhabited by more than 7 million people.

Fallout from Three Mile Island and Chernobyl

In 1979, the nuclear accident at Three Mile Island (TMI) near Harrisburg, Pennsylvania, created anxious times for the citizens of Sacramento. Rancho

Seco was a clone of the TMI reactor. Both were designed by Babcock and Wilcox (B&W), one of four U.S. reactor manufacturers. The NRC ordered the shutdown of all B&W reactors until they could determine what went wrong at TMI. The nuclear accident also marked the first large-scale involvement of the public in SMUD's decision making. Hundreds of people attended several SMUD board meetings, asking questions about the safety of Rancho Seco. Two SMUD board members requested that the NRC hold public hearings in Sacramento. Thirteen protesters were arrested for trespassing after jumping the gate at the entrance to Rancho Seco. Their trial ended in a hung jury.

One of the issues that surfaced at the trial was that radioactive iodine had been found in milk supplied to a local dairy from cows that grazed in pastures next to the plant. A high-level Rancho Seco manager stated that the contamination was a result of Chinese nuclear bomb tests that had occurred three years earlier. A local physics professor, Homer Ibser, calculated that the probability that atmospheric weapons tests had caused the radiation was less than one in a trillion, since the half-life of iodine-131 is only eight days. Ibser, who was involved with the Manhattan Project in 1942 and taught a class entitled "Living with Nuclear Energy" at the local university, was distressed that a utility executive would deliberately mislead the public about this important health issue.

Among those attending the SMUD board meetings in 1979 was Martha Ann Blackman, a 35-year-old mother and poet. She had linked up with a group of local anti-nuclear activists who became known as Sacramentans for Safe Energy or SAFE. In her soft-spoken manner, Blackman was to remain a thorn in SMUD's side for a decade to come. She and other members of SAFE would attend every SMUD board meeting and routinely ask questions about Rancho Seco during a period set aside for public comments. At the time, SMUD management saw Blackman and the safe-energy activists as nothing more than a minor inconvenience. The group had only a small following and attracted little attention from the press once the TMI accident faded from memory.

Although the performance of Rancho Seco began to deteriorate by 1982 it was assumed that the facility would continue to be SMUD's principal source of power well into the 21st century. However, in 1986 the nuclear power industry suffered a major blow to its credibility. On April 26, a major nuclear catastrophe occurred at one of the four Chernobyl reactors located just north of Kiev in the Ukraine. The extent of the radioactive contamination from Chernobyl was frightening, and people in Northern California began to buy iodine tablets as a precaution to protect themselves from thyroid cancer. The Chernobyl story was to dominate the news throughout the world for weeks. Anxiety about the safety of nuclear power increased in communities that were located near atomic reactors.

After Chernobyl, SAFE decided it would try to qualify a ballot initiative in Sacramento to close Rancho Seco. The first problem that the safe-energy activists faced was to figure out how to draft language that would stand up in court. Mike Remy, a local attorney who helped write California's Environmental Quality Act, began to sort through this thorny issue. Though he didn't consider himself to be against nuclear power, Remy had been concerned about the generation of large volumes of nuclear waste "without knowing what to do with it." Before Chernobyl, Remy had hoped for a reasonable dialog between those for and against nuclear power. He observed, "The two sides were not intellectually hooking into each other. They were trying to shock each other."

However, after the Chernobyl accident Remy decided that it would be necessary to confront the nuclear power industry head-on with a citizen's initiative. As a public power agency, in a state that permits citizen initiatives, SMUD provided an excellent setting for a debate about nuclear power. Remy and Blackman and the other safe-energy activists were motivated by a concern for nuclear safety. However, the SAFE initiative, as the anti–Rancho Seco ballot measure was called, focused on economics. Federal law preempted state or local governments from closing a nuclear power plant because of safety concerns. However, since the voters of Sacramento owned Rancho Seco, they could close the reactor for financial reasons. Although SAFE had not conducted an economic analysis of the consequences of shutting down Rancho Seco, they had the intuition that it might not be too costly. Their belief was founded principally on the plant's poor record of operations. By the time SAFE started circulating petitions to close the plant permanently, Rancho Seco had been out of operation for 11 months and SMUD had raised rates twice to pay for repairs.

The Role of the Local Media

The press played an important role in the debate over Rancho Seco. Both Sacramento newspapers, *The Sacramento Bee* and the *Sacramento Union*, assigned full-time reporters to the SMUD beat from 1985 through the two elections that included votes on Rancho Seco. In addition, Sacramento's major television station, KCRA (Channel 3), had a reporter who covered Rancho Seco issues for over a decade and had numerous contacts at the power plant. The major news radio station in Sacramento also assigned a reporter who covered SMUD for three years. It was not unusual for all four local television stations to send news crews to SMUD board meetings to cover controversies at the utility. Nonetheless, the most important part of the media for SMUD was *The Sacramento Bee*.

Since its inception, SMUD had always been able to count on the support of *The Sacramento Bee,* which is owned by the McClatchy family, long-time Sacramento residents. The McClatchy family had been strong supporters of public power and sided with SMUD in all its disputes with PG&E. The paper had also strongly supported SMUD's early involvement in nuclear power. But by 1986 the paper's editor, C. K. McClatchy, was beginning to have doubts about Rancho Seco. The Chernobyl accident increased the scrutiny *The Bee* gave to Rancho Seco.

In May 1986, *The Bee* ran a massive five-part series on Rancho Seco, entitled "Neglected Past/Uncertain Future." It portrayed a plant mired in trouble that had been shielded from the public eye for years. The articles pointed out that SMUD was run by a publicly elected board of directors, but the utility had been run "more like a private club than a public utility." In fact, until 1976 no incumbent on the board had lost an election. Typically, a member would resign in mid-term. The SMUD board would then install a new member who would be the incumbent in the next election. Not only was the board an "old boys club" but it routinely approved items presented to it by the utility's management. Few questioned this arrangement during times when electric rates were declining and there were no controversial issues. But the problems at Rancho Seco ended the days that decisions could be made without public involvement. As the media spotlighted who was on the SMUD board, the public began to realize they had a say in their utility's future.

The Bee noted that Rancho Seco had experienced three rapid cool downs that could threaten the reactor's integrity. The paper ran a vivid graphic portraying a reactor meltdown (see Figure 2-1). It showed a blob of radioactive material that had eaten through the plant's concrete foundation and was burrowing into the earth where it would remain highly radioactive for millennia.

The attention the media placed on problems at the nuclear plant prodded the NRC to look into past practices at Rancho Seco. Rancho Seco is unusual among U.S. nuclear reactors in that it is not located near a large body of water. It gets cooling water through a canal that links the plant to the Folsom Reservoir on the American River. Only a tiny creek drains out of the plant. It provides water for cattle on several ranches downstream before it empties into the Sacramento Delta. When the plant was built, a Rancho Seco manager told the local community that there would be zero releases of water contaminated by radioactive isotopes.

But as leaks developed in the plant's steam generators, SMUD was forced to break that promise. The NRC allowed SMUD to dump radioactive water into tiny Clay Creek, but the radioactive dosage could not exceed prescribed levels. In reports filed with the NRC, SMUD appeared never to exceed

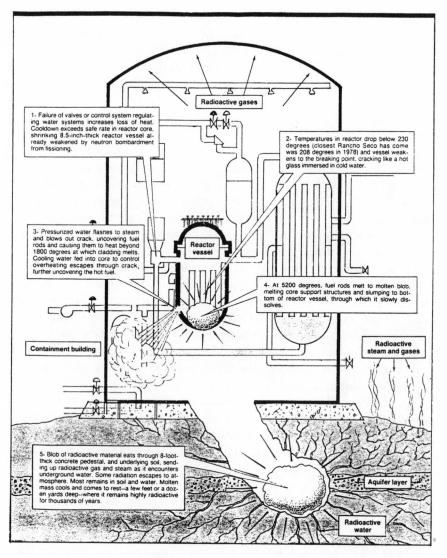

Figure 2-1: A nuclear reactor meltdown scenario. *Source: The Sacramento Bee.*

those limits. The NRC, however, suspected that SMUD had manipulated its radioactive sampling methods so it could discharge contaminated water that was overloading on-site storage. The NRC accused SMUD personnel of intentionally altering sampling methods. News of this alleged falsification of data triggered a $1 billion class action lawsuit against SMUD by ranchers and farmers downstream from the plant.

Seeking Out New Options

The first cracks in the community's support for the nuclear plant were beginning to appear in 1986. Elections for the SMUD board had spawned a community debate about alternatives to Rancho Seco. One of the nuclear plant's most avid supporters lost his seat to a newcomer who questioned the wisdom of additional investments in the nuclear power plant.

A week before the board elections, SAFE kicked off a campaign to give voters the final say on the troubled nuclear plant. The group would need to collect 25,000 signatures, a formidable task since SAFE did not have the resources to pay signature gatherers, the most common technique used in California to qualify initiatives. They went ahead anyway. The public's response to the initiative petition was astounding. Using only volunteers, SAFE collected and delivered 50,630 signatures to SMUD on April 27, 1987. This was more than twice the number needed to qualify their ballot measure. SAFE urged SMUD to put the measure on the ballot of November 3, 1987. SMUD wanted to put the election off until November 1988.

Meanwhile, Rancho Seco was taking its toll on Sacramento ratepayers. In two years, SMUD spent $410 million on Rancho Seco while paying out an additional $205 million to purchase replacement power.

By mid-1987 SMUD's rates were only 15 percent below those of neighboring PG&E. Another rate increase would put SMUD's rates for most residential customers above those of PG&E, undermining the rationale for public ownership. SMUD was also running out of cash. By the fall of 1987 the prospect of another rate increase forced the SMUD board to consider something it had long resisted doing—seriously looking at alternatives to Rancho Seco.

SMUD President Cliff Wilcox recommended that SMUD seek bids for energy as a possible replacement for Rancho Seco and simultaneously order management to develop a "wind down" plan. Wilcox said he had heard that there was at least 2,000 megawatts of excess energy in California. "There's no need for Rancho Seco if there's enough energy out there," he said. As a result of Wilcox's suggestion, SMUD issued a Request for Proposals (RFP) offering to buy power at prices below 4.2 cents per kilowatt hour. That was SMUD's estimate of its average cost for bulk power. If SMUD could buy power for less than that amount on the wholesale power market, it would not have to raise rates.

Competitive bidding for new power sources was initiated by the Maine Public Utilities Commission in 1984, but until the SMUD RFP it had not been done before by a utility on any significant scale. Power sales were usually arranged between utilities that were connected with each other. Long-term "integration agreements," like the one between SMUD and PG&E,

were the common mechanism used to coordinate power transfers between utilities. There was a small spot market for short-term electricity sales. It was mainly driven by hydroelectric production that during a rainy year might displace the use of more expensive fossil fuel power plants.

Most independent power projects stimulated by PURPA have been developed under standardized contracts that had been authorized by state public utility commissions. SMUD's proposal to bid for power supplies was different. It was a simple yet bold idea. The results of this competitive process helped open up the wholesale power market in the western United States. After SMUD, a number of other municipal utilities, including the giant Los Angeles Department of Water and Power, would conduct competitive bidding processes. Even the California Public Utilities Commission would experiment with a competitive process for developing new power sources as an alternative to standardized power contracts. However, none had as compelling a need to identify new sources of power as did SMUD.

Even though the SMUD board was beginning to look into alternatives to Rancho Seco, they were still very uncomfortable with the SAFE initiative. They refused to set a date for the initiative election, despite pleas from a crowd of over 200 people at one tense SMUD board meeting. The refusal forced SAFE to seek a court order to determine when the election would be held.

The majority of the SMUD board felt that the nuclear plant had value as an asset. However, several board members thought the plant was too financially risky for SMUD to run alone. They were hoping they could find a qualified utility to buy all or part of the plant. While the date of the election on Rancho Seco was still up in the air, SMUD sent out a prospectus offering to sell Rancho Seco. The price was negotiable. The most natural prospect would be PG&E, which was physically interconnected with SMUD.

PG&E decided they were not interested in operating Rancho Seco, but they did offer to buy SMUD outright and take responsibility for delivering electricity to Sacramento. They also promised to close Rancho Seco. In addition they would freeze rates for several years and hire all current SMUD employees. SMUD was, all of a sudden, in play. But, unlike a private corporation, there were no stocks for competing suitors to buy. But there were plenty of political and legal considerations. For one, a PG&E buyout would have to be approved by the California Public Utilities Commission.

As SMUD contemplated PG&E's offer and waited to see if anyone was interested in buying Rancho Seco, it was inundated by proposals for power sales from utilities and independent power producers. PG&E itself had offered to sell SMUD enough power to fully replace the output from Rancho Seco. In making this offer, PG&E said it was willing to negotiate the terms of a power sale with SMUD, but that the buyout of the municipal utility by PG&E offered the most "comprehensive benefits" for Sacramento.

The Bonneville Power Administration in the Pacific Northwest offered to sell 725 megawatts, roughly three-quarters of the capacity of Rancho Seco, at about 3 cents per kilowatt hour. Southern California Edison proposed to sell SMUD 400 megawatts of surplus capacity while an independent power developer wanted to convert Rancho Seco to operate on natural gas. Portland General Electric said it was interested in taking over SMUD's generation assets, including Rancho Seco, and selling power back to SMUD. All of a sudden, SMUD had many ways to meet the utility's future obligations.

PG&E's takeover proposal came under attack by Campaign California, a statewide activist group organized by then State Assemblyman Tom Hayden, the founder of Students for a Democratic Society. The group argued that a PG&E takeover would result in the cost overruns incurred at the utility's Diablo Canyon Nuclear Power Plant being passed on to Sacramento's ratepayers. They also criticized the further concentration of power into the hands of the nation's largest utility.

New Leadership for SMUD

As these events were unfolding, SMUD recruited Richard Byrne, who had headed up the Massachusetts Municipal Wholesale Electric Company, an agency that had its own problems with a nuclear power plant. The Massachusetts utility had been deeply involved in the construction of the Seabrook nuclear plant in New Hampshire.

Byrne came to Sacramento determined to turn around the deeply troubled municipal utility. His first act was to authorize an independent study of the options for SMUD's future. He decided that outside experts would perform this study since SMUD itself lacked credibility with both the public in Sacramento and the financial community. Merrill Lynch Inc., the New York investment banking firm, was selected to head the study group, which was to be called the QUEST team. QUEST was an acronym for Quality Energy for Sacramento's Tomorrow. Besides Merrill Lynch, members of the QUEST team included nationwide experts in the fields of engineering, economics, public policy, and resource planning. The QUEST team was expected to complete its study by February 1, 1988, a month after Rancho Seco was scheduled to start running.

With a new general manager, SMUD's relations with PG&E and other investor-owned utilities began to thaw. However, PG&E, which once enjoyed an exclusive arrangement with the Sacramento utility to provide bulk power, was put on notice by Byrne that it would have to compete to keep SMUD's business.

Byrne was beginning to surprise utility and community leaders alike. In November 1987, he hinted that Rancho Seco might not be essential to SMUD. Before a group of Sacramento businesses, Byrne announced that he was studying six alternative scenarios for SMUD. Besides the status quo and the PG&E takeover, they included converting Rancho Seco to run on natural gas, transferring the ownership of Rancho Seco to another utility, transferring all power plants to other utilities, and replacing the output of Rancho Seco with purchased power.

Byrne's openness to alternatives to Rancho Seco was new. For some, it was a breath of fresh air. For others, it was very worrisome. A small but powerful group of local businessmen representing large industrial customers and real estate interests had quietly been laying the political groundwork for the sale of Rancho Seco to Duke Power. Led by a prominent attorney, Joseph Coomes, they thought that SMUD was too political an organization to run a nuclear plant and wanted to turn the plant over to a private company. Duke operated similar nuclear plants in North Carolina and they ran better than Rancho Seco. Byrne's comments startled Coomes and his allies in the nuclear power industry.

Byrne, who had a reputation as being a supporter of nuclear power, welcomed Duke Power's overtures. However, he didn't consider the Duke takeover of Rancho Seco to be SMUD's only or best option. He wanted to understand the pros and cons of various options. Some of the members of the SMUD board, nonetheless, began to suspect that he might recommend the closure of the nuclear reactor. For these SMUD board members, as well as managers at the utility and a significant part of Sacramento's business community, running Rancho Seco had become the measure of SMUD's success as a utility. It was impossible for them to envision a viable municipal utility that did not own most of its own sources of power. Shutting down Rancho Seco would mean that SMUD would become dependent on others for power. In addition to the local concerns, the national nuclear power industry began to see that the closure of Rancho Seco might further undermine public support for nuclear power in the United States.

The Rancho Seco Campaign Begins

The date set for the election on the fate of Rancho Seco was June 7, 1988. Campaign California kicked off a door-to-door campaign to shut down the nuclear reactor with a flyer whose message was: "Rancho Seco: We cannot afford the risk." SAFE set a goal of raising $250,000 to finance radio and television advertisements.

PG&E decided to drop its bid to takeover SMUD after conducting a telephone opinion survey about its proposal. On the same day, Byrne revealed that SMUD was entering into a ten-year agreement with Southern California Edison to buy up to 700 MW of power at rates lower than those offered by PG&E.

A few days later, Byrne announced that he was close to an agreement with PG&E to purchase enough power to replace Rancho Seco. The new agreement would price power at the same level as the deal with SCE. Byrne also was waiting for information from Duke on the formation of a consortium of utilities to buy Rancho Seco. Duke said it was willing to buy 50 percent of the reactor and could line up another purchaser for an additional 25 percent. But, it still needed a West Coast utility to buy the remaining 25 percent.

On February 1, 1988, Byrne released the QUEST study. It found that there would be no significant difference in rates regardless of whether Rancho Seco was restarted or shut down. The study cautioned that any decision about the nuclear plant's future could not be made on the basis of economics alone but had to take into account perceptions of risk. Byrne said he would make a final recommendation on Rancho Seco after hearing from the public who were, after all, the owners of the reactor. The study narrowed down six possible scenarios to two—continued ownership and operation of Rancho Seco or its closure and replacement with power purchased from SCE and PG&E. The possibility of selling Rancho Seco had apparently fallen through since Duke Power could not line up a West Coast partner.

When SMUD held a public hearing on the QUEST report later that month, more than 50 speakers argued the pros and cons of Rancho Seco for 5 hours before an overflow audience. Following the testimony, Byrne and his panel of experts on the QUEST team recessed to a closed-door meeting. When he returned at 1:15 a.m., SMUD's general manager recommended closing Rancho Seco. Byrne said SMUD would benefit in the long run by replacing the large nuclear reactor with smaller power plants, by investing in energy conservation measures, and by striking deals to buy power from other utilities, moves which signaled a shift away from a centralized source of power toward more dispersed and diverse sources. The outside panel of legal, economic, engineering, and utility planning experts unanimously concurred with the recommendation, saying that continued operation of the plant was too risky to the financial well-being of the utility.

The Nuclear Industry Rallies

Byrne's verdict on Rancho Seco did not deter the plant's supporters, including Byrne's rival, Carl Andognini, who was overseeing the restart program at

Rancho Seco. Andognini was pushing to bring in Duke Power to run the plant. He warned that closing Rancho Seco would be a blow for the nuclear industry, even though the decision was based on economic considerations specific to SMUD. These remarks indicated a shift in the posture of the nuclear industry. A couple months earlier, the U.S. Council on Energy Awareness, a Washington, D.C.–based pro-nuclear lobbying organization, had announced that it had no plans to get involved in the Rancho Seco campaign. The group downplayed any repercussions that closure would have on the rest of the nuclear industry.

Andognini, however, saw the closing of Rancho Seco as impacting the viability of nuclear power in other parts of the nation. Two plants he mentioned were Seabrook in New Hampshire and Shoreham on Long Island. Both plants were having difficulties getting operating licenses.

Leaders in the nuclear industry began to rally behind Andognini's somber assessment. Duke Power owned three plants of similar design to Rancho Seco. A spokesman for the North Carolina company said that problems for one utility with a nuclear plant implied problems for others. Another Babcock and Wilcox (B&W) plant located near Toledo, Ohio, had been shut down for repairs and would be out for at least six months. The reactor manufacturer feared that with Three Mile Island already closed, another shutdown would create pressure to close the remaining six B&W plants.

Safe-energy activists saw the Rancho Seco situation in a similar light. Diane D'Arrigo of the Nuclear Information and Resource Service, based in Washington, D.C., said, "If this plant goes down for good it will be a very clear demonstration that nuclear power is not, and never was, an economically viable business." To the surprise of many, a Wall Street securities analyst also weighed in on the side of closing the plant. William Fish, who worked with Donaldson, Lufkin and Jenrette, a New York–based investment banking company, said his firm would upgrade its evaluation of SMUD's bonds if Byrne's recommendation to close the plant was approved. He argued that closing the reactor would eliminate the major source of credit instability for the utility.

Duke Power added a twist to the deliberations. It said it might be willing to run the plant if control were taken away from the SMUD board and given to a nonprofit corporation. What that meant was far from clear. Transferring a nuclear license to a new entity with no assets would certainly be a first for the NRC. Nonetheless, Duke said it was willing to enter into new negotiations with SMUD.

Rancho Seco employees had come early to a March 1988 SMUD board meeting, packing the meeting auditorium. They wanted the board to put a second measure on the June ballot. The tension was heavy both on the dais and in the audience. Four board members approved of putting a second referendum on the ballot that would allow Rancho Seco to operate for up to another 18 months. The new referendum called for a second vote on the nuclear plant after the trial run. The plant would be closed automatically if it ran below a 50 percent capacity factor for four consecutive months, unless four of the board members voted to keep it operating.

The June election could be decisive for the future of Rancho Seco. But, with two ballot measures there might be confusion. If both ballot measures passed, the one with the most votes would prevail. If both measures lost, then the future of Rancho Seco would once again be in the hands of the SMUD board.

As the campaigns for the two ballot measures got underway, the business leaders who had supported restarting the plant formed a political action committee to raise money. The group, calling itself Citizens for Affordable Energy, hired a campaign firm that had been involved in referenda on nuclear power plants in other states.

The pro-Seco effort would have ample fiscal resources. "Duke Power gives $100,000 in Seco fight" read a banner headline on the front page of *The Sacramento Bee* on April 6th. This generous campaign contribution was a wake-up call to those activists who thought they might have an advantage by using the initiative process to close Rancho Seco. Duke defended its contribution, claiming it was part of a larger effort to raise money to defeat the SAFE initiative, called Measure B, and support the SMUD referendum, labeled Measure C. A Duke spokesman explained that since the North Carolina company was going to operate Rancho Seco, they needed to explain who they were to the voters. The press, including *The Sacramento Bee,* took a less sanguine view of these campaign contributions (see Figure 2-2).

William Lee, Duke Power's Chairman of the Board, had been involved in nuclear power for most of his career as he rose through the ranks at the North Carolina utility. He had personally inspected the Three Mile Island plant after its partial meltdown. Following that accident he pushed for the creation of the Institute of Nuclear Power Operations, a utility-sponsored organization that evaluates nuclear operators, which he chaired until 1982.

Lee sent telegrams to other utilities with nuclear power plants requesting that they donate $10,000 to the campaign. He also came to Sacramento to tell the community why Duke was interested in their problem-plagued nuclear reactor. Duke wanted to operate the plant, he said, to prevent it from besmirching the reputation of nuclear power across the country. "We are hostage to the performance of every nuclear plant in the world," he said. He conceded that the North Carolina utility would not invest any money in Rancho Seco. He argued, just the same, that the utility was still at risk because if a problem developed at the plant the value of Duke's stock could decline.

While in Sacramento, Lee met with Byrne to try to conclude a draft Rancho Seco management agreement between the two utilities. A key stum-

'We've finally found a safe, clean, plentiful fuel for Rancho Seco!'

Figure 2-2: This cartoon captures the cynical views of the community regarding efforts to continue operating Rancho Seco. It also displays how, after many years of support for Rancho Seco, *The Sacramento Bee,* too, became disillusioned with SMUD's management of the reactor. *Source: The Sacramento Bee.*

bling block was what type of organization would act as a buffer between the SMUD board and Duke. Duke was adamant that it not have to report to a publicly elected board.

The news about the Duke–SMUD deal dominated both the written and electronic media and was changing the focus of the debate about Rancho Seco. Duke Power had become the issue. Opponents of Rancho Seco argued that Duke was getting a sweetheart deal. Supporters of the plant described Duke as the white knight SMUD needed to keep the nuclear reactor up and running. However, the Duke deal unraveled as soon as Byrne briefed individual board members about its terms and conditions. Duke was demanding a $25 million a year management fee plus a performance bonus that could give the company another $20 million a year. Duke would provide one senior plant manager and consultants as needed. SMUD would still be responsible for all ongoing costs at the plant. None of the five SMUD board members could support the deal. Rancho Seco's staunchest supporter, Ann Taylor, said Duke was asking too much for what it was giving. She said SMUD could run the plant itself.

The Campaign Heats Up

Rancho Seco's supporters, which included the local Chamber of Commerce, builders and realtors, farmers and construction trade unions, began running television ads in mid-April, six weeks before the election. With the early airing of the ads, SAFE knew they would be vastly outspent in the campaign. To get on the air they requested Sacramento's television stations to provide free time under the fairness doctrine, a policy of the Federal Communications Commission (FCC) to balance election-related programming.

Rancho Seco's supporters claimed that the fairness doctrine had been repealed by the FCC. An FCC official stated that a ruling in 1987 pertaining to general advertising in favor of nuclear power found that television stations did not have to run opposing ads. But, he said the decision did not rule out the use of the doctrine in an election. SAFE's attorneys pressed the TV stations and all but one eventually agreed to run anti–Rancho Seco ads without charging the group.

As the election neared, huge sums of money began to flow into the pro–Rancho Seco campaign, most of it from nuclear equipment suppliers, contractors, and other utilities that owned nuclear power plants. They outspent SAFE and Campaign California, the two organizations coordinating the shutdown campaign, by about five to one. With more than 250,000 people going to the polls on June 7, 1988, the trial run ballot measure passed 51.6 percent to 48.4 percent. The shutdown proposal lost 49.4 percent to 50.4 percent.

Shake-up at SMUD

The election gave Rancho Seco a reprieve. The trial run measure promised there would be a second election. Rancho Seco's performance needed to improve if voters were going to have confidence that it should be operated over the long haul.

Just a week after the voters had spoken, SMUD management announced that two rate increases of 8 percent each would be needed over the next 18 months, in part due to the Rancho Seco delays. Four days later, the SMUD board fired their general manager, Richard Byrne. The decision came during a six-hour closed-door session. Before returning to Massachusetts, Byrne revealed he had been "stifled, pressured and threatened" by pro–Rancho Seco board members. He felt the public was not well informed about the rate impacts of the trial run proposal.

However, soon after Byrne left, his work to establish better relationships with other California utilities began to pay off. SMUD signed a ten-year contract with Southern California Edison to buy up to 700 megawatts of power and agreed to buy between 400 and 1,000 megawatts from PG&E through 1999. SMUD had a year to determine exactly how much power it wanted to take from each utility. By testing the wholesale power market and encouraging competition between the state's two largest utilities, SMUD had figured out how to tap into the region's surplus power to meet its need for electricity without Rancho Seco. The power purchase contracts included off-ramps so SMUD could keep its future options open. This flexibility would prove valuable as competition in the wholesale power markets grew.

Andognini stepped down as the top manager at Rancho Seco in June 1988. The board gave him a $200,000 bonus for getting the plant restarted. Seven years after the nuclear plant was closed, Andognini still believes Sacramento made a mistake. He blamed the problems at Rancho Seco on SMUD for hiring too many consultants. "Every problem went before a committee or they found a consultant. Either way, nothing got done. And the consultants just ate away the money."

Byrne was given $320,000 in severance pay and was replaced by Dave Boggs who, until a few months earlier, was the head of Sacramento's regional transit agency. Well connected in Sacramento, Boggs had no experience working for an electric utility.

Boggs was hoping that SMUD had weathered the worst publicity with the firing of Byrne. He wanted to reduce the amount of attention SMUD got in the press. However, after the management turmoil the press intensified its coverage of SMUD. *The Sacramento Bee* assigned two full-time reporters to cover SMUD, including Tom Knudson, a Pulitzer prize–winning journalist that they had recruited from *The New York Times*.

Knudson loved to pry into obscure files. He planted himself at SMUD and began to read every document he could get his hands on. Among his discoveries were:

• SMUD had paid $458,000 in bonuses to Rancho Seco employees in less than two years.

• SMUD's tab for consultants at the plant was $102 million, some earning as much as $1,000 per day. He reported that, at times, consultants outnumbered employees at Rancho Seco.

• A Rancho Seco manager ordered $220,000 worth of blazers, shirts, trousers, and ties for the plant's employees to improve morale. Executives were to receive 2 blue blazers, 13 pairs of wool slacks, and 13 dress shirts. Not only was SMUD to purchase the clothes but it was to pay $130,000 to dry clean them for three years.

The public was incensed. Radio talk shows were jammed with irate callers. Letters to the editor poured into the two local newspapers.

A New Board and a Second Election

In November 1988, three new members would be elected to the SMUD board. Two had supported the closure of Rancho Seco and one had supported keeping it running. However, one of the new board members was having second thoughts about his earlier opposition to the plant.

As the new board was settling in, Rancho Seco experienced new troubles. Technicians had installed a governor that controlled an emergency feedwater pump—backwards. The pump sped out of control and overpressurized piping that fed water into the reactor core. The NRC immediately ordered SMUD not to touch any of the equipment without their approval. They wanted to find out what had happened and whether it might be repeated at other nuclear power plants.

A letter from INPO President Zack Pate suggested that Rancho Seco management may not have learned its lesson from the December 1985 overcooling incident. "Many aspects of the January 31, 1989, auxiliary feedwater pump failure, as well as a December 12, 1988, reactor scram that had caused one of the steam generators to dry-out event, are similar to the earlier serious events that led to the prolonged shutdown of Rancho Seco," Pate wrote.

Before this incident occurred, SMUD management had decided to call for a quick second election—June 6, 1989—just 12 months, instead of 18 months after the trial run measure had passed. The rationale for the early election was that it should occur before the plant needed to be refueled, a costly endeavor. Measure K, as the new referendum was called, was a simple up or down vote on the plant: a "yes" vote kept it open; a "no" vote shut it down.

This time around, SMUD employees dominated the campaign to save Rancho Seco. Campaign California managed a door-to-door canvas that continued through the election. SAFE organized phone banks to identify sympathetic voters. The 1989 campaign slogan was "Enough is Enough: Just Say No to Measure K."

On the tenth anniversary of the Three Mile Island accident, Rancho Seco broke down yet again. A reactor feedwater pump malfunctioned and caused pressure to increase in the reactor core. The plant then automatically shut down. Rancho Seco engineers were baffled by the latest problem. A senior plant engineer, in all seriousness, suggested that it might be due to sun spots. As SMUD was trying to figure out this latest glitch, Bechtel and B&W—firms which built the plant and supplied the reactor—announced they would share in Rancho Seco's future financial risk if Sacramento voted to keep the plant running.

As the election neared, Joe Buonaiuto, who had been elected board president with the help of two pro-Seco members, made up his mind to lead the campaign to keep Rancho Seco open. His first step was to support negotiating with Bechtel and B&W for a risk-sharing agreement for the future operation of Rancho Seco. At a contentious meeting the SMUD board voted 3-2 to request its staff to come back with a proposal by May 18, 1989, three weeks before the June 6th election date.

Buonaiuto proved to be a wild card for the proponents of Rancho Seco. Early in the campaign, he proclaimed ratepayers would be better off "unloading the plant for a dollar" than shutting it down. If a majority of ratepayers voted that SMUD could not operate the plant, Buonaiuto said he would try to find someone else that could. The widely publicized remarks stunned the community, most of which had assumed that the June vote would, once and for all, determine the fate of the plant. SAFE attorney Michael Remy said the SMUD board would face the immediate wrath of the voters if they ignored the outcome of the election.

Three weeks before the election, SMUD entered into a contract with Bechtel and B&W. The agreement would have SMUD pay the companies from $108 to $240 million for work at the plant over the next 5 1/2 years. The companies at the time received about $30 million in annual compensation. That could increase to $45 million if the plant ran well and could drop to $20 million if the plant ran poorly.

The arguments were now set for the election. Instead of Duke Power, the pro–Rancho Seco side was now claiming that Bechtel and Babcock and Wilcox would solve all of SMUD's problems. Rancho Seco's supporters didn't run any TV ads until the last two weeks of the campaign to prevent SAFE from asking for free time under the fairness doctrine, as they had in the previous campaign. Ads opposing Seco ran, therefore, as often as those that supported the nuclear plant. Instead of TV buys, Seco supporters put much of their money into a massive absentee ballot campaign. Plant workers went door to door identifying supporters of Rancho Seco and signing them up to vote by mail.

Rancho Seco's vote was drawing national attention. Closure of Rancho Seco "would send a message that, even though up to this point we have not been so successful at stopping a problem-plagued nuclear plant, it can be done," said Scott Denman of the Safe Energy Communication Council, a consortium of energy, environmental, and public interest groups. He added that he was very excited about Sacramento's ratepayers taking the cost issue into their own hands.

On June 6, 1989, a large voter turnout resulted in the rejection of Measure K by a 53.4 to 46.6 percent margin. Sacramento became the first community to close a nuclear reactor by public vote. The volunteers who had worked for over two years on the campaign felt they had made history and had regained control of their municipal utility. They had witnessed the dominant role that nuclear contractors and vendors had in shaping SMUD's policies for the past decade. They now hoped the utility would be able to find a better way to meet the community's energy needs. Some envisioned SMUD becoming a leader in the development of a safe, clean energy policy.

The vote hinged on a discussion of the economics of nuclear power. A majority of the voters were convinced that there were less risky and more cost-effective ways of meeting the community's power needs. The principal message of the vote was simple. Rancho Seco was no longer competitive and the community was better off without it.

Polls conducted by Campaign California showed the biggest opposition to Rancho Seco came from long-term residents. They had witnessed all of the problems, the endless promises, the huge rate hikes. According to Bob Mulholland, a Campaign California campaign strategist now employed by the California Democratic Party, the vote against Rancho Seco had "ramifications way beyond Sacramento." He observed, "If a coal or an oil plant can't compete, they are closed. Why should nuclear be different? Why try to protect these sacred cows?"

While the closure of Rancho Seco brought to an end a community debate about nuclear power in Sacramento, monumental tasks lay ahead. The safe lay up of Rancho Seco was the first priority. Then, SMUD had to figure out how to minimize the pain of the closure to the plant's workers. A decommissioning plan had to be developed. And new long-term power sources had to be identified.

Reactor for Sale

Although the plug was pulled on Rancho Seco on June 7, 1988, the majority of the SMUD board decided to see if anyone was interested in buying the nuclear reactor. A little-known, small San Francisco Bay Area company called Quadrex Corp. made an offer. The company's chairman William Derrickson announced he would like to restart the plant in November, but he would need to get NRC permission first. Derrickson had recently joined

Quadrex, a publicly traded company that provided consulting and low-level radioactive waste disposal services to the nuclear industry. It had working capital of less than $15 million and seemed an unlikely company to own and operate a nuclear power plant. Derrickson, however, had a reputation as a hotshot in the nuclear industry; completing Seabrook was his most cited achievement. *Forbes* called him the "Superman of the nuclear industry." The *Wall Street Journal* called him "a manager with chutzpah."

Derrickson's proposal to take over Rancho Seco got the attention of three national anti-nuclear groups—the Nuclear Information and Resource Service, the Union of Concerned Scientists, and the Safe Energy Communication Council. They announced that they would challenge any application to transfer a nuclear license from SMUD to Quadrex. They questioned Quadrex's financial and technical capabilities. Nevertheless, SMUD began negotiating with Quadrex to transfer the ownership of the plant.

Cliff Wilcox, the senior SMUD board member, was struggling with the proposal. He did not want to see more SMUD ratepayer money spent on the plant than would be necessary to close it. "The bottom line issue is whether Quadrex is big enough to absorb the costs whether the transfer be in six weeks, six months or six years," was the way he summed up the issue facing the SMUD board. To boost its credibility, Quadrex announced that the large engineering firm Stone & Webster had joined them in the negotiations to take over Rancho Seco. Details of their involvement were kept a secret.

Quadrex's final proposal was submitted to SMUD in September 1989 and called upon the utility to open and operate Rancho Seco while waiting for NRC approval of the license transfer. SMUD would then buy all the power from the plant at fixed prices that started at 5.95 cents per kilowatt hour. SMUD would also have to delay work on closing the plant until a proposed December voter referendum on the deal could be held.

Wilcox was leery of Quadrex's latest offer. He disliked the requirement that SMUD operate the plant until the nuclear license could be transferred. "What I see is a muddled program, commingled funds and trying to operate the plant with a team that has never operated the plant before." On September 11th—97 days after the election that closed Rancho Seco—Wilcox said he could not be convinced to support the Quadrex offer. "In my estimation it's not a deal that can fly." With that, the SMUD board voted unanimously to dump Quadrex and to permanently close Rancho Seco.

The closure of Rancho Seco was described by Scott Denman of the Safe Energy Communication Council as a "shot heard around the world." While the nuclear power industry had experienced the cancellation of orders for dozens of nuclear power plants since the accident at Three Mile Island, the closure of an operating nuclear plant challenged the assumption that once built nuclear plants were a cost-effective source of electricity.

Representatives of the nuclear power industry claimed that SMUD's situation was unique and that the closure of Rancho Seco did not mean that the continued operation of other reactors should be questioned. On the other hand, they also predicted that the decision to close an operating nuclear reactor would be disastrous for SMUD. In fact, many in the industry predicted the demise of SMUD as an independent utility. Some even attempted to make it difficult for SMUD to quickly reduce the work force at Rancho Seco by intervening before the Nuclear Regulatory Commission. Their goal was to make the closure as costly as possible while preserving the possibility that the plant might someday be restarted, if not by SMUD, then by someone else.

For both nuclear proponents and foes alike, how SMUD responded to the closure would be important in decisions that citizens made about nuclear power in other communities. *The New York Times* described the vote as "a popular repudiation of the conventional wisdom about nuclear plants: that they may be expensive to build but, once running, make electricity cheaply." Opponents of nuclear power saw the closure as a watershed event. SECC's Denman declared "Rancho Seco may well be remembered as the Waterloo of U.S. nuclear power development."

Setting a New Direction

Following the final closure of Rancho Seco, a new majority on the SMUD board of directors emerged. It was one that decided to put the interests of residential ratepayers first and make energy conservation a major focus of the organization. At the beginning of 1990, they elected a new SMUD board president who convinced the board that they needed to recruit a new general manager who agreed with their goals. After a national search the board finally hired S. David Freeman, a manager whose career earned him the nickname "utility repairman."

At age 64, Freeman was politically savvy and possessed a quick and biting wit. He was an immediate hit with the media since he was accessible and a master of the sound bite. Freeman also understood that SMUD's image needed a complete makeover. He had no doubt that he was the man to do it. Freeman pledged to be a consumer advocate for the ratepayers of Sacramento. "The people own this utility and instead of dividends they have been getting higher and higher bills," he said, promising to draw the line on rate increases. He also said he intended to mount an extensive energy conservation program.

Repairing SMUD's badly tarnished image would be no easy task. One state legislator was calling for the takeover of the municipal utility by the county. Another wanted state government oversight of SMUD. A coalition of right-wing groups led by the local chapter of Newt Gingrich's Conservative Opportunity Society was circulating a petition to place a privatization initiative on the November ballot. Several large industrial customers were suing SMUD over rates.

On the positive side, SMUD had finalized long-term power sale contracts with PG&E and SCE, assuring reliable supplies of electricity through 1999. Even lower cost power was becoming available from the Pacific Northwest and Canada—much of it produced by Columbia River hydroelectric facilities—which meant that SMUD's purchased power costs were declining. SMUD's financial situation was good. None of the bond rating agencies had lowered SMUD's credit rating following the closure of Rancho Seco.

Freeman hit the ground running in June. Early on he met with local political and business leaders, flew to Wall Street to meet with the financial community, and introduced himself to SMUD employees. He let it be known far and wide that his top priorities were to freeze rates and implement an ambitious energy conservation program. In short order Freeman was able to convince local legislators to drop proposals to reorganize SMUD. Business leaders were impressed with his energetic style but were taking a wait and see attitude. Many still believed that big rate increases loomed on the horizon. The campaign to privatize SMUD quickly ran out of steam. Short on signatures and cash, the backers of the privatization initiative fell apart.

Freeman would remain at SMUD for 3 ½ years. During that time his energy level never flagged. He was always looking for the next new thing. Most of the programs that he started now form the core of SMUD's energy strategy.

Competition and Conservation

Even before Freeman came to Sacramento, some important new programs were already underway. In March 1990 SMUD sent out a call for new sources of electricity. Request for Proposals went out to potential power generators throughout the western United States and Canada to sell SMUD electricity for both base load and peak use periods. Power sales, new power plants, and even the conversion of Rancho Seco to run on nonnuclear fuel were welcomed.

A bidders conference attracted power sellers from as far away as North Dakota and New Mexico. Companies wanting to produce power from resources such a geothermal, wind, solar, natural gas, and coal showed up. Notices of intent to submit bids indicated 12,000 megawatts of power would be offered to SMUD, more than 12 times the capacity of Rancho Seco.

SMUD intended to put together a "diversified portfolio" of power supplies. SMUD's power purchase contracts were tied to the price of natural gas. If gas prices went up, so would SMUD rates. By buying power from a variety of sources, with contracts beginning and expiring at different times, SMUD would be able to cushion itself from the effects of unexpected increases caused by one type of fuel or technology. A portfolio approach would allow SMUD to take electricity when it needed it most and not pay for extra capacity when it wasn't needed.

SMUD's new board majority had promised to make energy efficiency a top priority at SMUD. A new study showed that if Sacramentans implemented a comprehensive energy efficiency program, the area would need no more electricity generating capacity in 2010 than it did in 1990. The report, conducted by XENERGY of Oakland, California, showed that as much as 36 percent of SMUD's long-term energy needs could be met by replacing inefficient lights, motors, and appliances, saving the utility about $1 billion over the next 20 years. The report was the highlight of a day-long symposium that SMUD sponsored for the community in May 1990. Over 200 business and community leaders heard from nationally recognized leaders in energy conservation, like Amory Lovins and Art Rosenfeld. After the symposium, Sacramento attorney Cleve Livingston observed, "If we set our minds to it, Sacramento can become the energy capital of California, if not the nation."

Chapter 3

The Road to Recovery for the Sacramento Municipal Utility District

> "Assume discontinuity in our affairs, . . . and you threaten the
> authority of the holders of knowledge, of those in charge, of
> those in power."
>
> —Charles Handy, *The Age of Unreason*

Closing the Rancho Seco nuclear power plant was necessary to immediately stabilize electric rates in Sacramento. SMUD not only had to manage the closure of Rancho Seco but had to come up with a plan to assure that the community would have an affordable and reliable supply of electricity well into the future. The survival of the nation's third largest municipal utility would depend on reducing exposure to risks like it had faced with Rancho Seco.

SMUD was in a unique position to put into practice some of the new ideas about energy planning being discussed in the 1980s. Energy analysts like Amory Lovins at the Rocky Mountain Institute claimed that the potential for saving electricity from new energy efficiency technologies was immense. All that was needed was the will and the resources to make it happen. He argued that this "soft path" to energy security could best be achieved at the local level with bottom-up solutions. SMUD made an excellent laboratory for testing the potential for implementing energy efficiency measures. Following the closure of Rancho Seco, the municipal utility was dependent on power purchased from other utilities. The electricity saved from energy efficiency would mean less money exported out of the community.

Another idea that had been much discussed among energy policymakers was competitive bidding for new sources of power. California's vigorous implementation of PURPA had demonstrated that electric utilities were not the only institution that could effectively develop new sources of power. Still, the question remained: How to make best use of the technical and financial capabilities of independent power producers to more directly benefit retail consumers? Up to this time, the most common mechanism for stimulating nonutility power projects had been to mandate long-term contracts with prices that were set administratively. Many advocated a more market-oriented mechanism. SMUD had successfully used a competitive bidding and competitive negotiation process to nail down power contracts with other utilities to replace Rancho Seco's power. It decided to use these processes again to make independent power developers compete against one another for the right to develop new power plants.

Since SMUD needed to aggressively implement energy efficiency programs *and* acquire new supplies of electricity, it had the opportunity to blend these resources together in a way that would increase economic and environmental benefits for the Sacramento region. However, there were a number of variables to consider in estimating the benefits and many uncertainties. In an effort to balance benefits and risks, energy policymakers had proposed that utilities adopt integrated resource planning (IRP). The fundamental concept behind IRP is that utilities need to better understand the cost of all resource options that are needed to satisfy the demand for electric services. These options include new power plants, modifications of existing facilities, power purchases, energy efficiency and load management measures, improvements to transmission lines, and alternative rate structures.

Until it closed Rancho Seco and ended its "integration agreement" with PG&E, SMUD did not have to pay much attention to these complex IRP issues. For many years SMUD's principal concern was how to manage the sharp "needle" peak that occurred during the hottest days of the summer. But with the closure of Rancho Seco, and increased dependence on power purchased from other utilities, SMUD would have to take IRP much more seriously. No other utility had as urgent a reason to make it work as SMUD, which projected a need for as much as 1,400 MW of additional power sources within a decade.

One of the issues that is usually addressed in IRP is the value of resource diversity. The people at SMUD had learned the value of diversity from their experience with Rancho Seco. The challenge for SMUD was to invest in technologies that had the potential for future cost reductions while not overly burdening existing ratepayers with higher costs. Advocates of renewable energy technologies, like Don Aitken, former president of the American Solar Energy Society, had called upon utilities to support a policy of "sustained and orderly development" to accelerate the commercialization of promising new energy technologies. SMUD was to embrace this concept and put in place a comprehensive program to promote advanced and renewable energy technologies.

SMUD also recognized that it was essential to involve the community in the decisions about what new resources to acquire. The active participation of customers would be needed to build what SMUD called a "Conservation Power Plant."

A Conservation Power Plant

By 1990, SMUD already had over a decade of experience in implementing load management programs that shift the time of electricity consumption. The peak demand for electricity in Sacramento occurs during the hot summer months of June through September when air conditioning use is the highest. The difference in demand for electricity in Sacramento from a 95 degree day and a 105 degree day can be as much as 300 megawatts or about 15 percent of SMUD's peak load. But those last 300 megawatts of consumer demand will occur for fewer than 50 hours out of the year, or less than 1 percent of the time. Even arranging to buy power to meet these needs is expensive since the days that the peaks occur cannot be predicted far in advance.

SMUD management decided back in the late 1970s that it made more sense to implement load management programs with an emphasis on controlling the peak summer load. Its most effective program consisted of putting remotely controlled switches on air conditioners. Using a radio signal, SMUD could switch off air conditioners for periods of 20 minutes to four hours during a heat storm. Other programs were targeted to large industrial customers who could turn off major pieces of equipment on short notice. Companies that agree to curtail power are given discounts for the electricity they consume the remainder of the time.

After the closure of Rancho Seco, SMUD put together a plan to meet future energy needs that included the creation of a much wider variety of energy efficiency programs. The programs were targeted to all classes of customers, including homeowners, renters, small business owners, farms, and industries. Incentives were structured to improve the efficiency of new buildings, to encourage customers to replace inefficient appliances and equipment, and to stimulate cost-effective retrofits of existing homes, offices, and factories. SMUD's programs provide customers with information, technical assistance, financing, and rebates.

One of the first programs proposed was the planting of 500,000 trees by the year 2000. It is a key component of a "Conservation Power Plant," which would have to be built piece by piece in partnership with SMUD's customers over the next decade. Sacramento is a city of trees. Before air conditioning, shade was the only relief from the scorching summer sun. But once air conditioning became common in the 1950s, tree planting in new subdivisions slacked off. SMUD's tree planting initiative involved a partnership with a local nonprofit group called the Sacramento Tree Foundation, which had been formed in 1982 to stimulate citizen interest in expanding and taking care of Sacramento's urban forest. By 1995, an average of 140 trees a day were being planted by the SMUD–Tree Foundation partnership.

Customers are not charged directly for the trees. SMUD, instead, purchases the trees in bulk and charges for them through its electric rates just as it would for a new power plant. The Tree Foundation recruits customers to participate in the program, helps identify the proper sites and gives guidance on tree care. The customers plant and care for the trees on their property. Scientists at Lawrence Berkeley Laboratory found that in Sacramento mature trees can reduce the use of air conditioning by about 30 percent.

Another way to lower the demand for air conditioning is to reflect the heat back into the atmosphere. Modern urban areas are typically covered with dark surfaces and have less vegetation than natural habitats. These two factors cause urban areas to be hotter than the surrounding countryside by as much as 8°F. This phenomenon is called the "urban heat island effect." Dark roofs absorb much of the incoming solar radiation, raising temperatures in and around buildings. Shade from trees and white roofs, as well as light-colored pavement, can reduce the ambient temperature.[1]

SMUD is planning to stimulate a market for white roofs in the residential sector initially by providing rebates. Longer term, SMUD's role will become more informational. For the commercial sector SMUD will promote white surfaces through demonstration projects, customer education, and case studies.

Tree planting and reflective coating are just two measures in an ambitious energy efficiency strategy that includes 17 programs—from the installation of solar hot water heaters to rebates for efficient refrigerators. Many of the programs were first started in 1990 and have undergone revisions as technologies have changed and SMUD has learned how to better implement them. Over time, the tools used to promote reductions in demand for electricity shifted from rebates and direct installation to low-cost financing and education. By 1991, SMUD gained the distinction of devoting a greater share of its utility budget to energy efficiency programs than any other utility in the country (see Figure 3-1).

By the end of 1995 SMUD's demand-side management (DSM) programs had resulted in a peak demand reduction of 372 megawatts, which is the equivalent of about 40 percent of the capacity of the Rancho Seco nuclear power plant. Of that amount, 176 megawatts was obtained from load man-

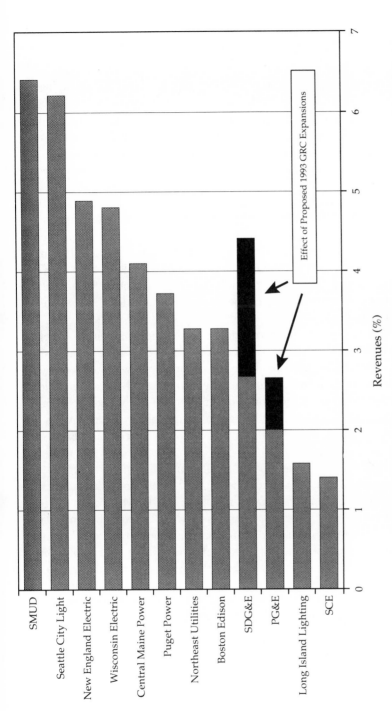

Figure 3-1: SMUD ranked at the top of the nation's utilities in earmarking a percentage of its utility revenue for energy efficiency programs in 1991, the year the utility began the full-scale implementation of its energy policy reform agenda. *Source:* ACEEE, "Increasing the Efficiency of Electricity Production and Use: Barriers and Strategies," November 1991, and NRDC updates.

agement programs that shift the time that energy is used and can be directly controlled by the utility. The remaining peak demand savings were obtained through reduction in energy consumption that resulted from the introduction of more efficient technologies like compact fluorescent lamps and high-efficiency air conditioners. In 1995 the energy efficiency measures that had been installed since 1991 reduced the consumption of electricity by 563 million kilowatt hours. At an average retail price of 7.7 cents per kilowatt hour, that meant that the electric bills for participating businesses and residents were about $43 million lower as result of the energy efficiency measures.

SMUD's energy efficiency programs have also helped to stimulate the local economy. A study conducted by the School of Business Administration at California State University, Sacramento, found that in 1992 SMUD's energy efficiency investments, $59 million on specific measures and $131 million in loans, had increased regional income by $124 million; created 878 jobs in the air conditioning, insulation, lighting, and other energy efficient-related industries in the Sacramento region; and added $22 million in wages to area households.

In addition, these programs have reduced adverse environmental impacts. By reducing the need for power production, SMUD's customers avoided significant amounts of air pollution. Even if compared to power produced by a new natural gas–fired combustion turbine with state-of-the-art emission control technology, the 563 million kilowatt hours saved through DSM programs reduced emissions of volatile organic compounds by 5 tons, oxides of nitrogen by 25 tons, carbon monoxide by 14 tons, and carbon dioxide by 265,000 tons.

These energy efficiency and load management programs have proven to be very popular with all classes of customers. By 1996 more than half of SMUD's customers had participated in at least one DSM program. These programs have helped make local businesses more competitive and have cleaned up Sacramento's air. For residential customers they have lowered monthly bills and created more of a sense of ownership of their local utility. A University of California, Davis, study of the tree program found that people who planted trees had stronger community ties, more interaction with their neighbors, and an overall greater satisfaction with their neighborhoods. The tree planting program illustrates how utilities using simple techniques can reduce the need for electricity.

Rate Reform

SMUD also looked to energy efficiency programs to correct historical inequities and inefficiencies in its rate structure. SMUD, like many utilities, aggressively promoted all-electric homes in the 1960s and early 1970s. It was assumed at the time that promoting the use of electricity for space heating and water heating would be cost-effective for the customer and would help balance the utility's summer and winter loads. Many homes in Sacramento were built with inefficient electric furnaces or heat pumps. Since an electric furnace uses a lot of electricity, SMUD provided winter discounts for all-electric homes. Still, when electric rates went up because of problems at Rancho Seco, these customers saw large increases in their monthly bills.

Despite the high bills that these all-electric customers pay, SMUD's cost of service studies showed that they were only paying about 85 percent of what it cost to provide them with electricity. Residential customers using natural gas for space heating were paying the full cost of service, while some large industries were paying 10 percent more than their share of costs. These large business customers, of course, objected to subsidizing other customers. However, if SMUD were to lower rates for industrial customers it would have to raise them for others. All-electric customers claimed that they had been enticed to purchase all electric homes by SMUD's special winter rates. They objected to proposals to end this discount.

SMUD's solution to this dilemma was to design an ambitious energy efficiency program for these all-electric homes. SMUD would provide incentives to replace inefficient electric furnaces with more efficient heat pumps, improve insulation, and replace electric water heaters with heat pump water heaters or solar water heating systems. Customers who agreed to participate in implementing these measures were then converted to the standard rate for electricity. This combination of incentives and price signals focused on improving the way the most inefficient households used electricity.

For industrial customers, SMUD offered time-of-use rates. These rates priced electricity at its marginal cost for different time periods. On-peak rates would be much higher than average while off-peak rates would be lower. This rate design encouraged companies to shift operations to off-peak time periods and to implement energy efficiency measures that would reduce peak consumption.

Competitive Bidding for New Power Supplies

SMUD anticipated that energy efficiency measures would meet much of the demand for electricity created by new growth, but it could not fill all of SMUD's power needs. The power purchase contracts with other utilities would eventually have to be replaced with new sources of electricity. Since SMUD is not supervised by the CPUC, it was free to develop a resource acquisition process that met its specific needs. SMUD's competitive bidding approach stands in stark contrast to a far more complex and ultimately un-successful attempt by the California Public Utilities Commission to use competition among independent power producers to add new power capac-ity to the state's system.

SMUD solicited bids for new power supplies for a second time in 1990. This time around, 65 companies submitted bids for 94 projects offering over 8,700 MW of power. SMUD needed just 850 MW to replace its contracts with SCE and PG&E. To screen offers, SMUD established five objectives to guide evaluation: minimization of price risks to maintain stable rates; minimization of environmental impacts, particularly air and water quality; reliability; stimulation of investment in the local community; and resource diversity. Bidders were instructed to provide information about the project cost, fuel source, operating characteristics, location, and the date the pro-ject would be on-line. SMUD provided information about the availability of transmission to applicants so that those costs could be estimated.

Guarantees on thermal efficiency became a major criteria for ranking pro-jects. This criterion advantaged the more advanced aeroderivative combus-tion turbines vis-à-vis the larger conventional industrial turbines since they use natural gas more efficiently. A risk evaluation of each proposed project was also performed. The risks SMUD looked into included the possibility of project cancellation, cost overruns, delays, deterioration in plant perfor-mance, the viability of the steam hosts, and the ability to obtain air pollu-tion offsets.

Finally, a public process was initiated to help determine how much value to place on the environmental benefits of proposed projects. Since the ratepayers of Sacramento owned SMUD it was essential that they deter-mine what mix of power plants best met their needs. The public was asked to evaluate four possible resource futures: a single large gas-fired unit at Rancho Seco; multiple local smaller gas-fired units; a resource mix that maximized renewable resources; and a resource mix that provided a diver-sity of fuel types.

Sacramento's top environmental concern is air pollution. The region's air quality ranks among the nation's ten worst. SMUD decided not to try to quantify the environmental costs of different types of air pollutants. Instead, it reported the quantity of pollutants associated with each power plant and resource scenario. Oral and written comments from the general public en-dorsed the diversity scenario. A poll of Sacramento County residents indi-cated strong support for renewable energy and a willingness to pay more for clean energy because of concerns over air quality.[2]

The public realized that the utility could not rely solely on renewable power plants since such power plants are more costly than fossil fuels, some are intermittent in nature (wind turbines typically only operate 30 percent of the time), and some tend to be located in remote regions that require greater investment in transmission infrastructure. In addition, SMUD required power plants that could be easily shut off and on in response to fluctuations in the demand for power placed on the system. For these reasons, SMUD chose the diversity scenario that included the construction of four small cogeneration power plants and a wind power plant. SMUD was later to add a biomass cogeneration plant located in the state of Washington to its resource portfolio.

Through a competitive bidding process, SMUD had identified four viable cogeneration projects in Sacramento. Two of the project developers were able to bring new host industries to Sacramento that could take advantage of large amounts of steam. One company was Glacier Valley Ice Co., which uses steam to drive refrigeration compressors in the manufacture of ice. The other is an ethanol production facility that uses steam to convert rice straw and other agricultural wastes into an alcohol that can be used as a fuel additive for gasoline.

Besides low-cost electricity, these projects bring other benefits to the community, including new jobs, a back-up power source to prevent sewage at the region's waste treatment facility from spilling into the Sacramento River, improvement in regional air quality, and a more reliable power grid. These cogeneration projects also enhance SMUD's competitiveness. Not only do they use natural gas very efficiently, but they help retain large industrial customers as ratepayers. In fact, one of SMUD's cogeneration projects was instrumental in convincing Campbell Soup to keep its 47-year-old food-processing plant in Sacramento and make substantial investments in its modernization. That decision saved over 2,000 union jobs.

Improving Air Quality

A major challenge for SMUD's cogeneration plants was compliance with federal, state, and local air pollution regulations. The federal Clean Air Act prohibits new power plants in a nonattainment region from adding to the area's air pollution. Every ton of pollutant emitted from a new facility must be "offset" by cleaning up an existing ton of local pollution.

SMUD worked closely with the Sacramento Metropolitan Air Quality Management District to implement a flexible, market-based approach to air quality improvement to allow for the siting of new power plants. SMUD paid a wood products manufacturer, an oil company, a dry cleaner, and a Sacramento River fish farm to upgrade their existing pollution controls or to switch from dirty diesel engines to electric motors. The reduction in air pollution from these measures created Emission Reduction Credits that allowed SMUD to site its new cogeneration plants.

SMUD is also helping to clean up the region's air by encouraging the use of electric transportation. Ninety percent of the air pollution in Sacramento can be traced to fossil-fuel driven transportation. Since 1990 the utility has promoted the use of light-rail, electric shuttles, and electric vehicles for personal and fleet use. SMUD efforts have focused on supporting advanced transportation technologies, stimulating local interest in electric transportation, and developing appropriate infrastructure for electric vehicles.

The utility has installed recharging stations at public parking lots, light-rail stations, at work sites, and in people's homes. A special time-of-use rate has also been adopted to encourage off-peak recharging.

Integrated Resource Planning

For many years resource planning at electric utilities was relatively straight-forward. Demand grew steadily. Technological developments and economies of scale lowered the cost of new power plants. Fuel prices and interest rates were stable. All these factors led to steady declines in electricity prices. Resource planning focused on the size, fuel, and timing of the next power plant the utility would build.

But by the 1980s a number of other factors began to be considered. They included greater access to transmission systems, innovations in demand-side management technologies, declining costs of renewable energy technologies, increased concern with environmental effects from electricity production, and considerable uncertainty about future load growth, the prices of fossil fuels, and the costs and construction times of power plants. These factors greatly complicated resource planning for SMUD and other utilities.

Different combinations of these resources are analyzed to see how well they meet future electric energy service needs and to determine their costs. The resource packages or portfolios are then subjected to an uncertainty or "what if" analysis. Different assumptions are made about local economic growth, future fossil fuel prices, and the cost of purchased power. A spectrum of estimates are used to evaluate the costs and performance of newer technologies like wind and solar power. Estimates of the environmental risks of various resources are also calculated.

Utilities report the results on alternative resource portfolios. Comparisons among portfolios show the types of trade-offs that must be made among competing objectives. The public can then evaluate how the utility's options reflect their values and concerns. Most advocates of IRP argue that a strong public participation component is essential in making decisions that embody community concerns. SMUD set an example of how to let its ratepayers voice their views on the overall direction the utility should take in the wake of the closure of Rancho Seco.

A key ingredient needed to make IRP work is for the utility to develop the proper tools for evaluating the cost-effectiveness of DSM programs. A number of tests of cost-effectiveness have been used that reflect different perspectives including those of participating customers, nonparticipating customers, the utility, all customers combined, and society as a whole. The test used most frequently by utilities to screen DSM programs has been the total resource cost test. That test looks at all the direct economic benefits of the energy efficiency measure and compares them to all the costs of implementing the measure, including those borne by the utility and by the participating customer. While the societal test is more comprehensive since it includes external benefits, it is used less frequently because of the difficulty of measuring those benefits.

In 1992 SMUD invited the Conservation Law Foundation (CLF) and the Natural Resources Defense Council (NRDC) to evaluate the utility's energy efficiency programs. CLF recognized that the approach SMUD initially adopted was appropriate for a rapid program start-up, but recommended that the utility adopt a more sophisticated screening tool to guide ongoing resource planning. To properly value efficiency savings, they recommended that SMUD adopt procedures that determine the value of energy savings at different times of the day and year, and for savings that would accrue over multiple years. SMUD now calculates the benefits of savings from energy efficiency programs based on the present value of avoided energy costs using multiple time periods (e.g., summer on-peak, winter off-peak) over the life of specific energy efficiency measures.

More recently, resource planners at SMUD and other utilities have also had to adapt to the introduction of more efficient combustion turbines and the decline in the price of natural gas. Until recently many utility planners had assumed that natural gas was in relatively short supply and that prices would increase more rapidly than the pace of general inflation. However, discoveries of new natural gas reserves, the development of new technologies for the exploration and recovery of gas, and an oversupply of pipeline capacity have kept prices low.

The preference that utilities had shown for investments in energy efficiency programs began to change with the emergence of modular, low-cost combustion turbines that could be installed with very short lead times and with the decline in uncertainty about future natural gas prices. To many utilities, large investments in conservation programs began to appear to be the riskier investment. New strategies will be needed to maintain the momentum these programs enjoyed in the early 1990s.

The change in natural gas prices has also affected the fortune of renewable energy technologies. Many utilities that several years ago expressed interest in renewable energy projects have since deferred them. In some cases the delays were caused by the uncertainty created by proposals to restructure the electric power industry. However, many were due to a growing gap in the price of electricity produced from natural gas–fired power plants and the price from renewable energy sources. It wasn't that renewable energy was becoming more expensive. Instead, gas prices were dropping and the efficiency of gas turbines was increasing.

Advanced and Renewable Technologies

In recognition of these barriers to integrating more renewable energy into the power grid, SMUD adopted a comprehensive strategy for acquiring renewable energy resources. It includes an all-renewable competitive solicitation that will allow the utility to acquire a modest amount of renewable energy—50 MW—in the near term and a longer term policy of supporting the commercialization of promising advanced and renewable technologies. In order to assess how these technologies can be integrated into a long-term resource portfolio, the SMUD board sets aside at least 1 percent of annual revenues to be used for an Advanced and Renewable Technology Development Program. This program allocates funds for the demonstration of a diverse array of promising technologies that offer environmental and economic benefits.

The goals of the Advanced and Renewable Technologies Development Program are to:

- Reduce the cost of selected technologies by investing in their accelerated development
- Promote early utility purchases of these technologies by identifying high-value applications
- Improve the reliability of the technologies through demonstration and testing
- Gain design and operating experience with selected technologies to prepare for commercial operation
- Promote significant technical improvements
- Facilitate integrated planning of advanced and renewable technologies

Since 1990 SMUD has made substantial in-roads in the deployment of three supply-side technologies that have the potential for widespread use: wind turbines, photovoltaics, and fuel cells.

Wind Turbines

By the time SMUD was ready to order a wind power plant, this source of clean power had already been widely deployed in California with impressive results. New wind projects had availability factors of over 95 percent. That compares very favorably with other power generation technologies. And the cost of building and operating a wind power plant was becoming competitive with other supply-side choices.

A new kind of wind turbine had been developed through a collaborative effort involving the Electric Power Research Institute, several U.S. utilities, and Kenetech Windpower, the largest U.S. wind developer. SMUD was the first utility to place an order for the new Kenetech Windpower variable-speed turbine. SMUD intended to build a 50-MW wind power plant but decided to proceed in stages to limit risks with the performance of this new type of wind turbine to its customers.

Wind power is particularly valuable for SMUD since it is routinely available at the times of year and day that SMUD needs the most electricity—late summer afternoons. The electricity generated by the wind therefore allows SMUD to avoid purchasing electricity from the wholesale market when demand is high, or it permits the utility to back down on its use of hydro power that can be stored for later use. The wind plant also provides a hedge against future fossil fuel hikes since the fuel is free. Though wind turbines do have some environmental impacts—particularly on the raptor population inhabiting the California Coastal Range—SMUD established a monitoring and mitigation program to ensure that its wind plant would not create any long-term impacts on local populations of birds at the Solano County site.

Photovoltaics

While SMUD's support for the wind project was a recognition that renewables can be deployed on a large-scale basis, SMUD was also interested in promoting smaller scale technologies that could be installed on the utility's distribution system close to the site of power consumption.

The best known of the modular electricity generating technologies that can be integrated into this vision of a "distributed utility" are photovoltaic (PV) cells that convert sunlight directly into electricity. In the early 1980s SMUD had built the largest centralized PV plant in the world—a 2-MW facility at the Rancho Seco site. But to make PV cells cost-effective for distributed utility applications, the industry needs to improve module performance, reduce manufacturing costs, and increase production volume.

SMUD is addressing these concerns through a program that is intended to attract investments in PV manufacturing to the Sacramento region.

In evaluating the cost-effectiveness of investments in photovoltaics, SMUD recognizes benefits that are not matched by conventional power technologies. Those benefits include:

- *Modularity.* PV systems can be installed in a wide variety of sizes. Utilities can gain experience with small affordable systems before making large-scale capital commitments.

- *Short lead times.* Both engineering design and construction times can be completed quickly once the decision to add photovoltaic capacity has been made. Capital is therefore not tied up for a long period and revenue can be generated as segments come on line.

- *Low operating costs.* PV systems have few moving parts. Photovoltaic cells and power conditioning units are solid-state devices. Thus availability is high and operation and maintenance costs are low.

- *Suitability for distributed siting.* Installation at strategic locations in the utility's distribution system can be preferable to transmission and distribution upgrades.

- *Minimal environmental impact.* PV installations do not produce emissions and require no water.

SMUD's commitment to commercializing photovoltaics helped stimulate the formation of the Utility Photovoltaic Group, which now is made up of over 90 U.S. and Canadian utilities. This organization has developed a multi-year program to stimulate the installation of grid-connected photovoltaics. SMUD's approach to the introduction of PV in Sacramento is viewed as a model in the Pacific Northwest and other regions of the country because of its use of "green pricing," a policy whereby citizens volunteer to pay a little extra for clean power.

Since 1993, SMUD has installed 3 megawatts of PVs in Sacramento, much of it on customers' rooftops. Customers who agree to let SMUD install PV systems on their property are called "PV Pioneers." SMUD installs and owns the systems but the customer agrees to pay higher rates (15 percent) in order to help stimulate the market for this environmentally benign technology. By 1996 SMUD had installed over 350 rooftop photovoltaic systems. Larger installations have been installed at the utility's distribution substations and as shading for parking lots.

More multiple-purpose PV projects are being planned at SMUD, including several that integrate PVs into building materials. SMUD's "green pricing" program is being examined by more and more utilities in light of the strong public support for renewable energy. It has the potential to stimulate even more investments in renewable energy technologies in a restructured electric power industry.

Fuel Cells

Another electric generation technology that holds much promise for distributed applications is the fuel cell. A fuel cell generates electricity and heat by combining a fuel and oxygen in an electrochemical reaction. It is more like a battery than a conventional power plant. Thermal power plants burn a fuel to create heat. The heat is used to produce mechanical energy in a turbine. The mechanical energy is then converted to electricity by spinning magnets through a wire coil. Fuel cells avoid these steps to produce electricity. In a fuel cell, hydrogen and oxygen are combined chemically to produce electricity, heat, and water (see Figure 3-2).

Fuel cells are capable of high efficiencies, low emissions, and capital costs that are competitive with new thermal power plants. The advantages of fuel cells are that they can be designed to follow electric loads with fast response time without significant losses in efficiency. This is a characteris-

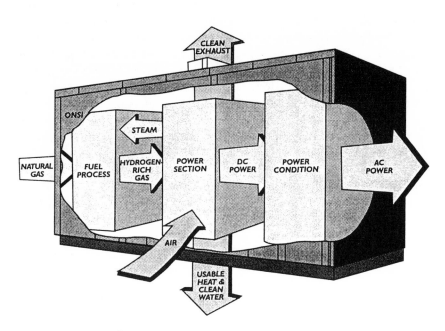

Figure 3-2: How a fuel cell works. *Source:* Sacramento Municipal Utility District.

tic that cannot be duplicated by a thermal power plant. It makes fuel cells particularly valuable for distributed applications where fluctuations in power demand occur frequently. Fuel cells are also cleaner than thermal power plants. A fuel cell using natural gas as its source of hydrogen will produce about half the carbon-dioxide emissions of a natural gas–fired power plant and 99 percent fewer emissions of oxides of nitrogen, a precursor of smog.

Both PV and fuel cells are technologies that can transform the economics of the power grid. They provide a set of benefits that are different from large-scale power plants. Utilities are still attempting to quantify the full spectrum of benefits that modular technologies like fuel cells can provide. Among the benefits for a utility are their high availability, lower reserve requirements, less reliance on peaking units, improved system reliability, improved power quality, faster return of capital, and ease of siting.

SMUD has already installed two 200-kilowatt fuel cells in Sacramento and is investigating several additional sites. The first unit was installed at a local hospital. The hot water that is produced as a by-product is integrated into the hospital's hot water supply, thereby reducing its consumption of natural gas. A second unit is located next to SMUD's headquarters in a location that is readily accessible to the public. SMUD is currently considering rewiring the unit to provide an uninterrupted power supply to an important computer system. During normal operations the fuel cell will provide electricity for the computer network. In the event the fuel cell shuts down a very rapid switch will connect the system to a circuit providing grid power. Other high-value applications are under investigation by SMUD, including highly variable loads like advanced medical imaging technologies and the area's light-rail system.

Independent Power Development

SMUD's location in the capital of the nation's most populous state has meant that the programs and policies it has adopted are highly visible to state legislators, the CEC and CPUC, and the executives of other utilities in the state. SMUD has also learned from California's early experiments in promoting innovative energy policies. One of the critical factors in SMUD's renaissance was the way the CPUC implemented PURPA, which engendered the growth of a robust independent power industry in the state. Developers of wind, solar thermal, geothermal, biomass, and cogeneration projects have invested about $13 billion in the state's economy. As SMUD put together its resource plan to replace Rancho Seco, it was able to tap into a reservoir of knowledge and experience with smaller scale, nonutility power plants that existed throughout California.

Most independent power plants have been developed using conventional project financing. After securing a contract to sell power to a utility, a project developer usually obtains funds to build the power plant through private capital markets. SMUD discovered after soliciting bids for new power sources that this method of financing would raise the cost of the projects by over 10 percent in comparison with tax-exempt municipal financing. However, issuing utility revenue bonds to build these projects would put SMUD's ratepayers at risk if the projects were delayed or canceled. SMUD was determined to avoid exposing its customers to additional financial risk after its experience with Rancho Seco. It was able to come up with a financing plan that reduced the power costs while keeping the development risks on the private entrepreneurs.

SMUD decided to establish Joint Power Agencies (JPAs) for its cogeneration projects that would allow it to use tax-exempt financing without putting its ratepayers at risk if problems developed at those facilities. SMUD entered into take-and-pay contracts with the JPAs to buy electricity. If the projects were not completed or did not perform as promised, the bondholders would be at risk, not SMUD's ratepayers. Of course, this meant the projects were a more risky investment for the bond buyers. However, the technologies were proven and SMUD had brought in highly qualified construction managers to build the power plants. This combination gave the financial community enough confidence that Wall Street's major credit rating agencies gave the projects a financial grade rating which meant that large financial institutions like pension funds and insurance companies would invest in the projects.

The only guarantee that SMUD made on behalf of its ratepayers was that it would purchase electricity from the power plants at agreed upon prices. The independent power developers were responsible for the licensing and permitting of the new power plants. Once completed the projects would be turned over to the JPAs and the developers would be paid a fee for their work and the risks they had assumed to develop and license the projects. If these projects perform better than expected, the JPA will return excess revenues to SMUD, which will allow it to lower rates or avoid future rate increases. SMUD was the first municipal utility to set up this type of JPA financing. This innovative arrangement is now being used as a model by other municipal utilities as well as government-owned utilities overseas (see Figure 3-3).

While SMUD was involving itself in innovative partnerships with entrepreneurs in the private sector to develop new power sources, California

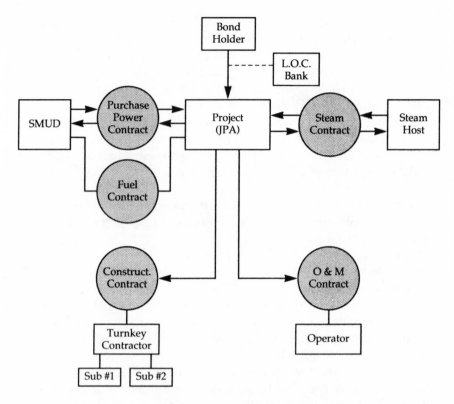

Figure 3-3: This diagram outlines the complex relationships between SMUD, the financial community, and private sector participants, including bond holders (and letter-of-credit banks), the industrial steam host where the facility is located, the construction contractor, and the operator of the power plant. The five key contracts that enable such a financing scheme to work are highlighted. *Source:* Sacramento Municipal Utility District.

state regulators were also looking at ways that the state's investor-owned utilities could use competitive markets to meet future power needs while avoiding risks to the state's electric ratepayers and improving the environment.

Past efforts to promote independent energy projects had contributed to increases in the rates charged by California's investor-owned utilities. But they had also assured that the state would not experience power shortages at a time when there was great uncertainty about the completion of several West Coast nuclear power plants. Before the Diablo Canyon, San Onofre, and Palo Verde plants, came on line in the mid-1980s, no one complained much about the cost of independent power projects. But after the nuclear plants went on-line and their costs were added to the rate base, concern about California's higher than average electric rates increased.

However, with the defense industry booming and jobs abundant, high electricity costs were accepted as part of the cost of doing business in California. That ended with the collapse of the Soviet Union and the end of the Cold War. California's economy, which had previously benefited from large defense expenditures, underwent a wrenching restructuring as the defense budget was cut. Jobs and economic development became important political issues. Industries began to complain about the state's high electric rates, which were 50 percent above the national average and far higher than rates charged in many neighboring states. California's Republican Governor Pete Wilson wanted to address this problem.

Wilson's appointees to the CPUC were very frustrated with the complex and mind-numbing processes that had been developed over the years to tinker with utility rates. CPUC proceedings were becoming frustrating exercises. The most notorious example was known as the Biennial Resource Plan Update (BRPU). Although it was called "biennial," skirmishing over the first and last BRPU lasted over eight years. The intent of the BRPU was to use a competitive bidding process for the acquisition of new power supplies for the state's investor-owned utilities. Its implementation became tied up in contentious and time-consuming regulatory processes and high-stakes politics.

Ironically, competitive bidding was initially embraced by state regulators, investor-owned utilities, independent power producers, and environmentalists after SMUD had shown how effective it was for replacing the power from Rancho Seco. However, as the BRPU process dragged on, the utilities decided that they did not need the additional power that would be developed through the bidding process. An increasingly competitive wholesale power market had made low-cost power from other parts of the West available to California's utilities. Since new power sources were not urgently needed, the investor-owned utilities had every incentive to delay the implementation of the BRPU and eventually kill it.

However, the BRPU allowed energy policymakers to better understand the environmental impacts of different types of power plants. Two state agencies spent countless hours over the course of several years computing the external costs of different air pollutants emitted by power plants. Coming up with a dollar value for specific pollutants proved controversial. Some advocated that it should be based on the costs associated with controlling air emissions. Others argued that it should be based on the actual impacts of pollutants on human health and the environment. A compromise was finally reached that allowed for the use of "clean air" bonus payments and penalties as part of the competitive bid process.[3]

The results of the BRPU competitive bidding process were impressive. Bids from independent power producers were 17 to 44 percent cheaper than utility proposals. Winning bids submitted to SCE offered prices far below the utility's nuclear as well as fossil fuel plants (see Figure 3-4). However, by the time the bids were received Southern California Edison decided that the projects were unneeded and would just add to their already high electric rates. SCE and the Independent Energy Producers tried to engage the public in the debate over the BRPU with an expensive public relations campaign that included full page ads in the *Wall Street Journal* and the *Los Angeles Times*.

Independent project developers were able to get a handful of Republican legislators whose districts would benefit from the new projects to put pressure on the state's utility commissioners. They finally convinced the CPUC

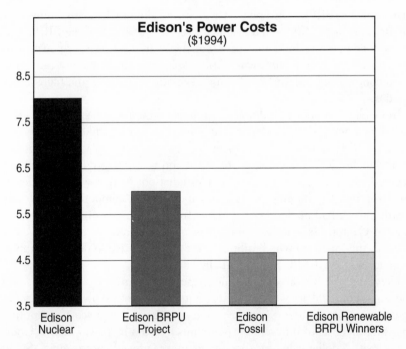

Figure 3-4: Regardless of the problems associated with the BRPU, this bar chart demonstrates that winning BRPU bids offered prices that look reasonable in light of SCE's own existing and proposed generation sources. *Source:* Independent Energy Producers.

to approve the BRPU, ordering utilities to enter into contracts with the private developers to build the new power plants. However, SCE did not give up on derailing the BRPU. The utility appealed the CPUC's decision before the Federal Energy Regulatory Commission (FERC), which oversees the implementation of PURPA. In February 1995, FERC issued an advisory opinion indicating that California's approach to competitive bidding was inconsistent with PURPA. Supporters of competitive bidding in California were flabbergasted by the FERC vote.

The complexity and the failure of the BRPU forced energy policymakers in California to rethink how new power resources should be acquired. It had become clear that a government-supervised competitive bidding process depended upon a consensus of all the parties involved, particularly the regulated utilities. Without such a consensus, or a sense of urgency that new power plants were needed, it became possible for one party to undermine the process. Of course, from SCE's point of view, the power simply was not needed and since no other power plants were to be closed to lower costs, rates would inevitably have to go up. Nonetheless, for California's independent power developers the process was devastating. Years of planning and engineering for new power plants had gone down the drain. Some renewable energy firms like the California-based wind developer Kenetech were financially weakened by the cancellation of the BRPU projects. Even though the costs of renewable sources had declined rapidly within a very short time frame (see Figure 3-5), companies that were world leaders in renewable technology development were suffering.

SMUD and California's Electricity Markets

Over seven years have passed since the voters of Sacramento closed their single largest source of power, Rancho Seco. During that time period, SMUD procured cost-effective sources of power to replace its nuclear reactor by using an innovative competitive bidding process and by instituting a variety of new energy efficiency programs. The municipal utility also completed the construction of a state-of-the-art cogeneration plant, with two more underway. Seventeen wind turbines were erected. Hundreds of PV systems had been placed on customers' rooftops and at other sites in the SMUD grid. SMUD became widely recognized as a national leader in the commercialization of renewable energy technologies. The major Wall Street bond rating agencies upgraded SMUD's credit rating three times. And since 1990, SMUD has not had a rate increase.

Much of SMUD's success has been due to its ability to take advantage of the opening up of the wholesale power market in the western United States.

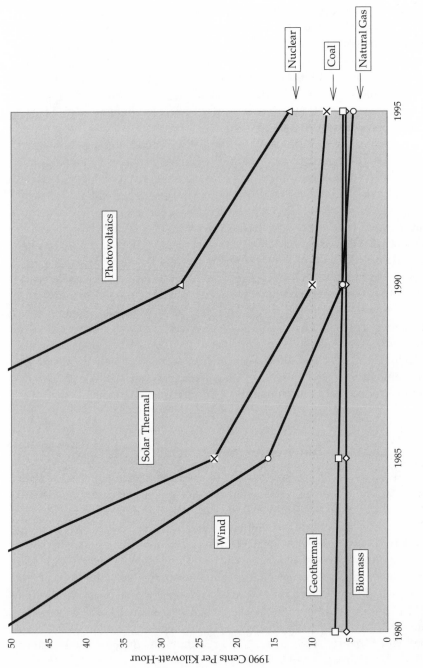

Figure 3-5: Renewable energy costs are dropping. *Source:* DOE, Energy Foundation.

The Western grid ties together the electric systems in 13 states (from New Mexico to Washington) and 2 Canadian provinces. Electricity moves through this grid at the speed of light, and thousands of power transactions are conducted daily.

Currently, the Western grid has a very large surplus of power supplies. In a competitive market, supply and demand would come into balance and the most expensive sources of electricity, which could not earn a profit, would be shut down. The electricity market is not yet fully competitive, however. Many utilities and nonutility power suppliers have been able to obtain government protection for their inefficient power plants and are able to pass on their fixed costs to captive customers.

SMUD has been the principal beneficiary of the surplus power market in the western United States. As the first utility in the West to close a nuclear power plant, SMUD has been able to shop for power for the past six years. Currently, SMUD buys almost half of the electricity it provides to its customers through the spot electricity market. Because of the power surplus, prices available in this buyer's market have been quite low. SMUD rates are now the lowest among major California utilities. For example, the municipal utility's 7.7 cents/kWh average rate is 23 percent lower than that of PG&E (9.9 cents/kWh), which serves the majority of Northern California customers.

This West Coast surplus of power also undermined the BRPU competitive bidding process. Despite the positive features of the new power plants—including new jobs and cleaner air—the utilities were unwilling to accept the risks associated with new long-term contracts. And while the proposed facilities were less costly than what the utilities had proposed building, unlike the situation at SMUD, there was no urgency to acquire new supply-side resources. This allowed California's investor-owned utilities to stretch out an integrated resource planning exercise that many had hoped to be a model of how to structure a competitive and environmentally responsible solicitation for new power sources. The experience of the BRPU revealed that if new and cleaner generation technologies are to be integrated into the power grid, it will be necessary to remove obsolete, uneconomic, and more polluting power plants from the resource base. The BRPU process had shown that it would be difficult for new technologies to find a place in a market with embedded subsidies for inefficient nuclear and fossil fuel power plants.

While SMUD has taken advantage of the current imbalance between supply and demand for power, it recognizes that the surplus power market is not going to last forever. Sooner or later older and less-efficient power plants will be closed by competitive forces. As supply and demand come into balance, the price for power during peak periods of the year will rise.

This will create opportunities for solar and other renewable technologies to enter the market.

SMUD's success, and the failure of the BRPU, set the stage for a reconsideration of the roles of California's energy regulators and the state's investor-owned utilities. The changes sweeping the region's wholesale power market were being noticed by the state's large industrial customers who sensed the time was right for proposing a restructuring of the power industry that would allow them to get access to lower cost power that utilities like SMUD were buying.

Notes

1. SMUD has carried out a number of experiments to determine the amount of energy savings from white roofs. Two school buildings were retrofitted with white roofs, which reduced the energy consumed by air conditioning by 35 percent. SMUD has also sponsored research on light-colored pavement. The research shows that the reflectivity of pavement can be improved by as much as 50 percent through various methods, including use of sand or oyster shells as a coating for asphalt. Potential applications include streets, parking lots, and driveways.

2. "Support For Renewable Energy is High in Sacramento County," Barry Sussman & Associates, 1919 Pennsylvania Ave. NW, Suite 300, Washington, D.C. 20006, Spring 1990, p. 28.

3. These externality values for SCE and SDG&E were $28,524 per ton of NO_x; $21,306 per ton of sulfur dioxide (SO_2); $6,171 per ton of particulate matter (PM); $20,374 per ton of reactive organic gas (ROG); and $30 per ton for carbon. For PG&E in Northern California the costs were: $8,272 per ton of NO_x; $4,060 per ton of SO_2; $2,380 per ton of PM; $1,180 per ton of ROG; and $30 per ton of carbon.

Chapter 4

The Breakup of Utilities
and California's New Electric Order

"Every abuse ought to be reformed, unless the reform is more
dangerous than the abuse itself."

—Voltaire

On April 20, 1994, the California Public Utilities Commission (CPUC)
shook up the nation's electric utility industry by releasing a proposal to restructure the nation's largest and most complicated electricity market. The
proposed policy was articulated in a document entitled "A Vision for the Future of California's Electric Services Industry," often called the "Blue Book"
because of the color of its cover. The CPUC proposal called for ending the
monopoly power of the state's investor-owned utilities, eliminating government supervision of resource planning for new power supplies, and giving
consumers a choice of electricity suppliers. The CPUC claimed that these
reforms would lower the state's electric rates.

The increasingly competitive wholesale power market in the western
United States, and the clamor by some large industries for lower rates, convinced the CPUC that they needed to find ways to promote more competition in the electric services industry. The BRPU process had been frustrating for the CPUC. Nevertheless, it proved that new power plants could be
developed readily at competitive prices. It also showed just how cumber-
some regulatory planning processes had become in a fast-changing electricity marketplace.

In stark contrast to California's state regulatory process, SMUD had competitively bid and licensed five power plants and had already constructed two in less time than it took the CPUC to finish the BRPU. It was easy to see that California's utility regulatory system needed reform badly.

Past approaches to utility resource planning were bumping up against the realities of regional wholesale markets dominated by power surpluses. The electric utility industry uses only about half of all the generating capacity that has been built in the United States. That means that fully staffed power plants are being paid for by captive ratepayers, yet remain idle for half the time. Some units have run less than 10 percent of the time since they were built. Part of the reason for this surplus is the fact that electricity cannot be stored in large quantities in an economic manner. Much of the surplus, however, is a legacy of cost-plus regulation for utilities that had always enjoyed monopoly status in their exclusive franchise areas.

The decision of the CPUC to reshape California's electric utility industry came as a surprise to many. Although the CPUC had previously held public hearings on industry restructuring, these hearings were viewed as more of an academic exercise than as laying the groundwork for a specific proposal. At those hearings, some academics and consultants recommended radical surgery on the existing electric utility industry. Utility representatives, however, argued that the structure of the industry was not a problem. Their preference was for more incentives, or what they called performance-based ratemaking.

The "Blue Book" called for the phase-in of direct customer access to power suppliers, or what had been called "retail wheeling." The CPUC claimed that this would lower rates by allowing customers to choose their power supplier. They also promised to protect customers who would continue taking electric service from their current utility supplier. The proposal also called for protecting utility shareholders through imposition of a surcharge (later called a Competitive Transition Charge, or CTC) that would prevent uneconomic assets, such as utility nuclear plants, from being abandoned or "stranded" by those customers choosing to buy power from lower cost power producers.

The proposal created headlines across the state. Stories were peppered with fine-sounding phrases like "consumer choice" and "competition." Upon closer inspection, the promises didn't appear to add up. Observers were puzzled by how the CPUC could lower costs for large industrial customers, protect small businesses and residential customers from rate increases, and guarantee that utility shareholder dividends wouldn't take a hit. They promised there would be no losers. Since the public had not been involved in developing the CPUC proposal, representatives of consumers were very suspicious. Groups like Toward Utility Rate Normalization (TURN) suspected that the CPUC was trying to shift costs to others with less political or market power.

The president of the CPUC, Daniel Fessler, set an extraordinary timetable for making this proposal the law of the land, at least in California. A final order would be issued in four months. Written comments from the public were due in 30 days. It wasn't clear whether the CPUC would hold any public hearings. The first customers were scheduled to have direct access to alternative power suppliers by the beginning of 1996.

The "Blue Book" and its peremptory schedule triggered a public outcry that the CPUC had apparently not expected. The stock value of California's three big electric utilities had fallen by a total of $1.7 billion one week after the announcement. Thousands of calls from irate shareholders poured into the state's utilities.

The strongest supporter of the CPUC proposal was the California Large Energy Consumers Association. CLECA was formed in the early 1980s to lobby the CPUC on rate matters. It is composed of cement companies, steel manufacturers, and a gold mining firm. The goal of CLECA has been to allow its members to bypass the state's utilities and get lower cost electricity. Despite their advocacy for more than a decade, CLECA's members had not been able to tap into alternative suppliers of electricity.

But CLECA had been successful in getting the CPUC to focus on industrial rate issues. All twelve CLECA members are now served by special discounted rates as low as 4.5 cents/kWh, less than half the rate paid by residential and small business customers. From 1985 through 1994, residential rates had increased 36 percent for PG&E and SCE. Over the same time period, industrial rates fell by 16 percent, a trend justified on the grounds that such changes reflected the cost of providing electric service (see Figure 4-1).

While large industries were pleased with the CPUC's bold reform proposal, independent energy producers were worried that it would end incentives for the development of renewable and other cleaner energy supplies. Environmentalists saw the plan as disastrous for their efforts to keep the state's utilities involved in promoting energy efficiency measures.

However, the issue that would be most controversial was the proposal to allow utilities to recover "stranded costs," that is, their investments in uneconomic power plants.

California's Stranded Costs

The term "stranded costs" is an unusual concept. Businesses operating in free and competitive markets do not have the luxury of identifying and col-

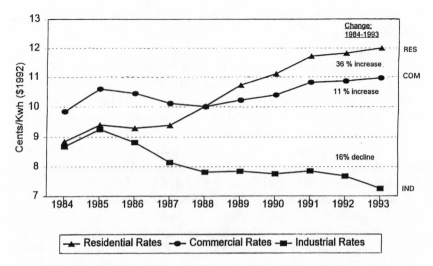

Figure 4-1: Trends in California electricity rates for investor-owned utilities by customer class, 1984–1993. (Compiled by Käri Smith, Resource Specialist, Natural Resources Defense Council.) *Source:* U.S. Department of Energy.

lecting such costs. There are only profits and losses. Among state and federal regulators of electric utilities, however, the phrase "stranded costs" has gained buzz-word status. The handling of stranded costs has been the most difficult issue associated with restructuring electric utilities. That's because the issue is inescapably linked with electric rates. If a goal of restructuring is to lower electric rates in the near future, then it is impossible to recover all costs left stranded by a competitive market.

As of 1991, California's rates were the fourth highest in the nation, with the average cost being just under 10 cents/kWh (see Figure 4-2). Severin Borenstein, director of the University of California Energy Institute, argued that there are two fundamental reasons for high rates in California: endowments and policy decisions. Endowments are things such as climate and the resource base—things that dictate the scope of energy options available. Policy decisions are discretionary and can be controlled by regulators and lawmakers.

Borenstein claimed that though some policy decisions made in the past may have turned out to be mistakes, that did not mean they were bad decisions at the time. The three policy choices principally responsible for California's rate dilemma were:

• the decision to invest in nuclear power in the 1960s and 1970s;

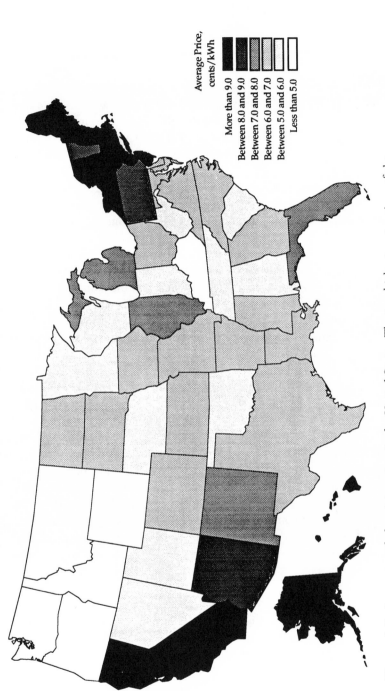

Figure 4-2: Average retail electricity prices in the United States. The two highest cost regions of the country are California and New England; the lowest cost region is the Pacific Northwest. *Source:* U.S. Department of Energy, Energy Information Administration, *Electric Sales and Revenue 1992*, Tables 8 and 9. DOE/EIA-0540(92), April 1994.

Average Price, cents/kWh

More than 9.0
Between 8.0 and 9.0
Between 7.0 and 8.0
Between 6.0 and 7.0
Between 5.0 and 6.0
Less than 5.0

- the decision to aggressively implement PURPA in the 1980s;
- the decision by the CPUC to stick with command-and-control/rate-of-return regulation.

Borenstein told the California Legislature, "The strongest link among areas with high electricity prices seems to be the decision to invest heavily in nuclear power . . . It is probably the case that regulatory oversight in California actually prevented an even larger commitment to nuclear power in the state."[1]

Utilities owning nuclear plants pointed to California's large independent power industry as the primary culprit responsible for higher rates. This debate about the cause of high rates is important in framing policy solutions that will lower the rates over time. The arguments are cluttered with competing claims and confusing numbers.

One fact is clear, however. Power generated by nuclear power plants represents a far larger portion of the energy consumed in the state. Therefore, its contribution to the overall California rate base is larger.

Robert Kinosian, a rate analyst with the CPUC Division of Ratepayer Advocates (DRA), argues that nuclear plants are indeed the largest factor in California's higher rates, but independent power projects with long-term contracts put in place in the early 1980s are also expensive. The prices utilities pay for energy under those contracts were linked to forecasts made at a time when rapidly rising oil prices were predicted. Some projects received payments as high as 13 cents/kWh. But the higher prices were only guaranteed for 10 years. In year 11 the price for energy drops to the utility's short-term marginal cost of electricity—that is, the price utilities pay for power in the wholesale power market. By the mid-1990s most independent power contracts had reached the "Year 11" price cliff. So, while independent power plants had been expensive, that problem would be partly resolved by the change in terms of their contracts with utilities. Of course, this did not correct the problem of surplus power that had been created by the construction of these facilities. The relative contributions of nuclear power and independently owned power plants to California's high rates are illustrated in Figure 4-3.

The question of which technologies led to high electric rates in California is clouded by special deals that the CPUC authorized for the state's nuclear power plants. The first deal was struck with PG&E in 1988 for its two-unit Diablo Canyon nuclear power plant. Ironically, it was SMUD's problems with Rancho Seco that figured centrally in the first-of-its-kind settlement between the CPUC and PG&E over a nuclear facility that had cost $5.5 billion to build. Under that arrangement Diablo Canyon's revenues are

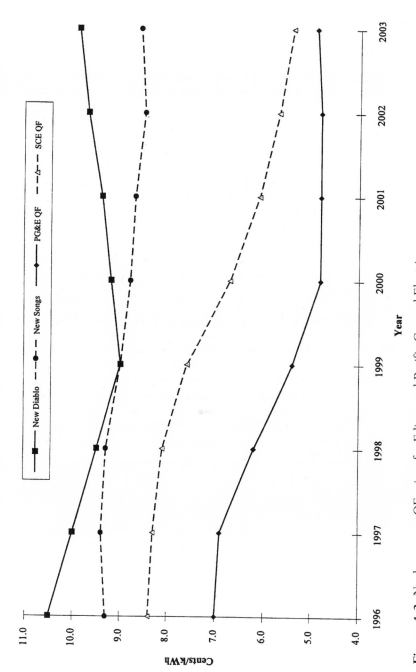

Figure 4-3: Nuclear versus QF prices for Edison and Pacific Gas and Electric. *Source:* CPUC Division of Ratepayer Advocates.

linked to its performance. The reason for the unusual settlement was to protect PG&E's ratepayers from future costs at Diablo Canyon. Under the settlement the utility and its shareholders are responsible for all future repair work or additional investments at the plant. In return, PG&E was given the opportunity to earn more money if the plant ran well. One of the chief negotiators for the state, Michael Strumwasser, said, "The only reason we fixed on this course was because of the Rancho Secos of this world." PG&E was to be paid a price per kilowatt hour for the power the plant produced. If Diablo Canyon ran well, the price PG&E could collect would go up. If it operated poorly, the price would go down.

At the time, this was a radical departure from the conventional method of setting utility rates. For all of PG&E's other power plants the CPUC exhaustively reviews all the prudent costs the utility claims to have incurred to build and operate them. This never-ending bean counting keeps a lot of people busy at both the utility and the CPUC. In fact, it is the primary function of PUCs across the nation. PG&E wanted to avoid the embarrassment of having the CPUC investigate all the mistakes and cost overruns that went into the building of Diablo Canyon. Therefore, it was willing to accept this novel agreement.

The settlement pegged the price of Diablo Canyon power to the projection of future fossil fuel prices. At the time the CPUC still thought that the price of oil and gas would rise rapidly in the 1990s. Since the settlement, Diablo Canyon has run reliably. Under the incentive-based rates adopted by the CPUC, PG&E was earning more than 12 cents/kWh for Diablo Canyon power in 1994. Power available in the West Coast wholesale power market was selling at or below 3 cents at the time.

PG&E itself recognized that these high prices were not sustainable and at the end of 1994 proposed to cut the price of Diablo Canyon's electricity by 32 percent over the next five years. More recently PG&E has proposed to allow Diablo Canyon to be subject to market prices in 2002.

SCE, too, put forward in 1994 a plan to prepare its San Onofre nuclear power plants for the harsher realities of a cost-driven electricity marketplace. Consideration of the SCE plan for the San Onofre plants got caught up in the debate over restructuring as discussed later in this chapter.

Once competition is introduced into the electricity market, costly facilities like nuclear power plants may not be able to earn a price that would compensate their utility/owners for prior investments. Likewise, the market price for electricity may be lower than the prices established in the standardized power purchase contracts for independent power producers. The difference between the market price for electricity and the revenue stream needed to pay off these facilities and contracts constitutes the "stranded costs." Nationwide stranded costs have been estimated to be as high as

$250 billion. California's share is estimated to be between $8 and $32 billion. Columbia University professor Bernard Black calculates that utility bills would drop by about 15 percent on the average—if these stranded costs were absorbed by the marketplace.[2]

It can be argued that the investments in SMUD's Rancho Seco plant were stranded when the voters of Sacramento decided to close the facility. However, there was no question who would pay for the stranded costs since SMUD had no shareholders. SMUD classified the abandoned plant as a regulatory asset and through its ratepayers will collect (through the year 2008) the funds needed to pay off the investment in the plant. Since it has proven less costly to buy power in the wholesale power market and to build more efficient cogeneration power plants and invest in energy efficiency, SMUD's ratepayers are better off even though they have to pay off the debt on Rancho Seco.

The need to deal with utility shareholders complicates the debate over what to do with uneconomic power plants. For ratepayers it may make sense to close the facilities and buy lower cost power elsewhere. For utility shareholders that means lost profits that they believe were promised to them by state regulators. Worse, it may lower the value of the utility's stock.

A Public Dialog Begins

In response to the controversy generated by its "retail wheeling" proposal, the CPUC agreed to hold a series of public hearings throughout the state. In September 1994 a standing room only crowd packed a hearing in South Lake Tahoe. They had come from up and down the High Sierra mountain ranges to discuss the restructuring of electric utilities.

Most of the audience that evening wore flannel shirts. Old and young were packed into the tiny city council chambers to tell representatives of the CPUC what they thought about the proposal to revamp the way electricity was produced, bought, and sold in California. At a press conference before the Lake Tahoe meeting, Tom Infusio, general counsel with the environmental group Friends Aware of Wildlife Needs (FAWN), stood together with Larry Lloyd, a logger. Earlier in the year FAWN had been hung in effigy by loggers in a fight over timber sales. Now they were working together because of mutual concern about more severe forest fires, contaminated drinking water, more crowded landfills, and unemployment.

They, and others, claimed that biomass power plants helped prevent forest fires because operators would pay people to collect underbrush. According to the U.S. Forest Service, 750,000 tons of dead and dying trees were removed from National Forests and converted to electricity in 1993. The Forest Service, in written testimony, warned that the CPUC proposal to deregulate electricity would "increase the risk of catastrophic wildfires." Robert Meacher, a Plumas County supervisor, pointed out that the biomass industry "defrays some of the taxpayer's costs relating to fire-proofing California's forests." This is an arena in which taxpayer and ratepayer interests converge. He warned that if the biomass power industry were harmed during industry restructuring that "the taxpayer will be forced to shoulder the full costs of forest fire-proofing."

Local governments worried about meeting state mandates for waste diversion. Air quality regulators were concerned about a return to open field burning of potential biomass fuel.

The fate of biomass plants was just one example of how the existing utility structure uses ratepayer funds to support programs that offer value to communities throughout California. But as the price for electricity in the wholesale market plunged to 3 cents/kWh or less, biomass plants accustomed to being supported by power purchase prices above 10 cents/kWh were scrambling to stay in business.

The mobilization of opposition to the CPUC proposal was the work of a broad coalition of groups. It involved groups representing residential customers, low income groups, energy service companies, renewable energy entrepreneurs, utility workers unions, and environmental groups. The list of parties intervening in this CPUC proceeding had grown to include over 500 names.

Not all businesses thought that the primary goal of reform of electric utilities should be to reduce rates. A national organization that represents some progressive businesses—the Business Social Responsibility (BSR) Fund— joined forces with environmentalists and representatives of residential consumers to criticize the CPUC's plan. The group argued that a competitive market for electricity should be created to provide customers with the highest quality and most reliable service. For companies that relied upon computers and other high-tech tools, an outage of a few milliseconds could cause very costly damages. The group encouraged the development of modular technologies that could be installed on-site to increase service reliability. An analysis performed for the Natural Resources Defense Council revealed that the industries pushing for retail wheeling did not represent the bulk of California's business community.[3]

The vociferous public opposition took its toll on individual commissioners. Two stepped down, leaving the CPUC shorthanded. Two were persuaded that the retail wheeling proposal was a mistake. Only one commissioner continued to support the original restructuring plan. Since the commissioners did not see eye to eye on any single plan, utility reform would have to wait until the governor appointed a new commissioner in April 1995.

A Wholesale Power Pool

Academics and other utility experts had also joined the debate over utility restructuring. One of the most influential was William Hogan, a Harvard professor who advocated creation of a statewide wholesale power pool as an alternative to the direct access proposal. He argued that the key ingredients to reform were an efficient wholesale market and performance-based regulation for the remaining elements of the utility monopoly. He believed that a wholesale power pool would provide all customers, large and small, with an equal opportunity to benefit from a competitive market.

Hogan's position was supported by NRDC. Their energy advocate, Ralph Cavanagh, believed that a wholesale pool was compatible with incentives that promoted cost-effective energy efficiency and renewable energy investments. He was particularly concerned that power suppliers not be rewarded for promoting the sale of electricity. He wanted to keep in place California's policy that "decoupled" utilities' profits from their energy sales. Others weren't as impressed by the idea of an efficient wholesale power pool. Leonard S. Hyman, author of *America's Electric Utilities: Past, Present and Future*, advised the CPUC to look at the flaws in the power pool put in place in England. He claimed that the pool was obsolete once it was created. He believed that an efficient market required that buyers and sellers interact in real time, rather than after the fact.

A revised proposal was finally released by the CPUC in May 1995. The new proposal dropped direct access in favor of wholesale power pool dubbed POOLCO, a reference to "pool company." Those opposed to the CPUC's original retail wheeling proposal breathed a sigh of relief. The new CPUC proposal called for setting up a spot market for the buying and selling of bulk power. Generators would submit bids to the pool in specific time increments, probably on a half-hourly basis. Buyers would submit demand bids in the same time increments. With this data in hand, the operator of the power pool would post a market clearing or "spot" price once supply met demand. All successful bidders would receive this uniform price for their power.

The state's three investor-owned utilities would be required to make all of their sales and purchases through this central pool. The CPUC hoped that municipal utilities such as SMUD would voluntarily participate in the pool. Existing contracts with independent power producers and wholesale power purchase contracts, as well as utility-owned nuclear and hydroelectric plants, would be exempt from bidding into the pool and would be scheduled first. All other power suppliers would compete in the competitive bidding process.

The state's two southern California utilities, municipal utilities, labor unions, representatives of minority consumers, and NRDC endorsed the concept of a more open wholesale power market. Groups such as the American Wind Energy Association were pleased to see that the CPUC had endorsed a "renewables portfolio standard," a concept the group had been pushing to support clean power options in a restructured utility world.

But as the groups looked into the details of the POOLCO proposal, some had second thoughts. Municipal utilities weren't sure whether POOLCO was an all or nothing proposition. They preferred a wholesale market that allowed for both a pool and bilateral transactions. Consumer groups worried about the concentration of market power that the three utilities would have in a unregulated wholesale power market. Many environmentalists saw the plan as exempting nuclear power plants from the forces of the market. They also worried that fully depreciated fossil fuel units would beat newer and cleaner power plants in being dispatched by the pool. Independent power developers also viewed the ground rules for the pool as tilted too heavily toward utility sources of power.

PG&E still preferred direct access over POOLCO, as did large industrial customers. Some environmental organizations, including the Sierra Club and the Environmental Defense Fund, began to question whether renewable energy would be better off under the POOLCO approach, even with a renewable portfolio standard. Rich Ferguson, national energy policy director for the Sierra Club, thought that if residential consumers had a choice about their supplier of power, more might opt for renewable energy sources. He argued that an approach suggested by the residential ratepayer group TURN might be better than POOLCO. TURN's proposal was called "community access," and it suggested that cities and counties act on behalf of their businesses and residents in purchasing bulk supplies of electricity. Electric service would still be delivered by the company that owned the wires. The management of the resource portfolio would be done by a publicly accountable body. This type of approach was a way for citizens served by investor-owned utilities to mimic SMUD's more diversified approach to acquiring new power sources.

Some environmentalists saw the community access approach as a way to let some communities choose to pay more for cleaner energy. Polling data showed that consumers preferred renewable energy—even if it cost a little more. Still, it was an untested idea. However, growing numbers of policy analysts supported using direct access as a development strategy for renewable energy.[4]

Opposition to POOLCO was growing. Some of the state's largest industries, oil companies, school districts, and the state agency responsible for buying electricity for state buildings joined together with advocates for residential ratepayers and some environmental advocates to oppose the CPUC plan. The state's attorney general, Dan Lungren, also warned that POOLCO could violate California's anti-trust statutes and invite price gouging by the state's investor-owned utilities, which owned most of the sources of electricity in the state.

Governor Pete Wilson was distressed that two of his important political allies, the state's investor-owned utilities and large manufacturing firms, were fighting. He was further stung by an op-ed in the *Wall Street Journal* that criticized the POOLCO proposal at the same time he was campaigning for the Republican Presidential nomination. His office quietly encouraged representatives of SCE and the California Manufacturers Association (CMA), the state's largest industry lobbying group, to talk. Other representatives of big power users and the Independent Energy Producers also joined these behind-the-scenes meetings. In August 1995 they reached an agreement.

The agreement, presented as a Memorandum of Understanding (MOU), blended a wholesale pool proposal with a direct access regime. It also brought to an end the dispute between the large industrial customers and SCE[5] over the proposed San Onofre Nuclear Generating Station (SONGS) settlement.

Industrial consumers had initially joined environmentalists in condemning the settlement that allowed SCE to accelerate its recovery of its investment in the nuclear plant. The MOU agreement would allow SCE to protect its investment in the nuclear reactor in exchange for allowing these industries direct access to power supplies. Independent power producers would get protection for their existing power purchase contracts, and they hoped to generate new business through direct access transactions.

In response to the MOU, environmental and consumer advocates developed their own list of principles to guide restructuring; it was called the "Framework for Restructuring in the Public Interest."[6]

The CPUC Decides

A political compromise shaped California's latest restructuring proposal. Achieving a consensus among commissioners regarding which of the two models of reform was better—direct access or a wholesale power pool—had become impossible.

At its last meeting of 1995, a divided CPUC adopted a plan based largely on the MOU offered by SCE, large industrial customers, and power plant developers. The new CPUC plan was a hybrid wholesale power pool/direct access model. Before the 3-2 vote in favor of his proposal, CPUC President Dan Fessler boasted: "With the decision today we open the door to the broadest array of choice in which former ratepayers can function as informed customers. The variety of customer choice which we seek for Californians after January 1, 1998 is currently unavailable anywhere in the world." He called upon the Federal Energy Regulatory Commission (FERC) to help shape an era of "cooperative federalism" to set the foundation for "a California consensus."

The CPUC's voluminous decision called for breaking up California's three big investor-owned utility monopolies—PG&E, SCE, and SDG&E— companies that employ tens of thousands of people, have tens of billions of dollars of assets, and serve over 20 million customers. Instead of three huge monopolies that manage the infrastructure associated with the generation, transmission, and distribution of electricity, the CPUC came up with a very different structure for California's $20 billion power market. The CPUC's final proposal was a compromise that promised to offer the benefits of competition to all classes of utility customers. New policies designed to support energy conservation and renewable resources—a systems benefit charge and a renewables portfolio standard—were also included in the blueprint for retooling the electric services industry.

In two and one-half years the CPUC had shifted from a direct access "retail wheeling" scheme to a mandatory wholesale power pool to a blend of the two approaches. Figure 4-4 is a graphic representation of the monopoly-based system as well as the hybrid proposal that would serve as the basic structure for California's new electricity market.

The new California market structure would consist of a wholesale power pool, called the Power Exchange, that would be open to all suppliers of electricity; a limited number of customers would, simultaneously, be granted the right to enter into direct access contracts. The Power Exchange would set and publicize the prices that would be paid for bulk power. A nonprofit company called the Independent System Operator, or ISO, would operate the high-voltage transmission lines that crisscross California and would be responsible for coordinating and scheduling the delivery of electricity over the transmission system. Local utilities would still be responsible for maintaining and operating the lower voltage power lines that are needed to deliver power to consumers, be they residential, agricultural, or industrial. Distribution would remain a monopoly service subject to state regulation, the last remnant of today's existing monopoly utilities.

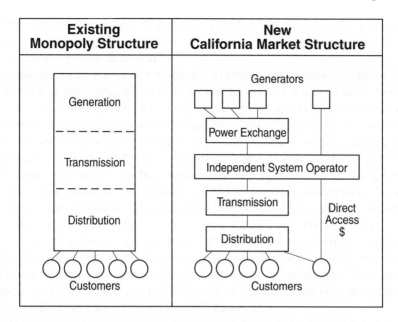

Figure 4-4: Today's market structure is based on vertically integrated utility monopolies. California's new market structure will feature a system in which most of the state's electricity will flow through a power pool called the Power Exchange. A new entity—the Independent System Operator (ISO)—will schedule transmission service but transmission and distribution will be provided by distinct companies. While the financial flows in direct access transactions are separate from the Power Exchange, direct access purchases will still rely on the ISO and transmission and distribution infrastructure. *Source:* Energy Foundation, CAS Graphics.

The Power Exchange

The Power Exchange will be a statewide spot energy market somewhat like the New York Stock Exchange. Every day sellers will submit bids for the sale of electricity on a half-hourly basis for scheduling the following day. The purpose of the exchange would be to set a clearing price for bulk power for each period of time. Bids that beat the market clearing price would have their power plants scheduled to operate the following day. Information about the spot market price would be made available to investors (as well as other interested parties) to help them determine when and where to build new electric generation facilities.

The spot market price would allow consumers who have special meters to make decisions about their electricity consumption in real time. Customers could curtail consumption during periods of high prices and shift demand to off-peak periods, when prices are low. Generators of power would be able to evaluate cost-cutting steps that might be needed to make their units competitive. The spot market would also prompt the creation of a risk-hedging market that would allow consumers to enter into so-called "contracts for differences" that would give them longer term price predictability.

The FERC will have the final say over the Power Exchange since it oversees wholesale power sales according to the Federal Power Act. Nonetheless, the CPUC felt it would be useful to enunciate a set of principles that should be used in designing the operation of such an exchange.

It was of fundamental importance that the Power Exchange have no interest in any source of power generation so that it could make unbiased decisions about determining the spot market price. Nondiscriminatory bidding rules would be established. Power plants would be ranked based on the costs identified through bids. This ranking would facilitate scheduling of power delivery by the ISO.

The CPUC decided to require that the state's three investor-owned utilities bid all of their power generation into the exchange for the first five years. This was done to simplify the task of determining the value of stranded assets. It was also assumed that this policy would protect consumers who would continue to rely on an existing utility for the full range of electric services. By preventing utilities from immediately entering into direct access contracts, a right granted to all other sellers of power, the CPUC thought these utilities would have an incentive to divest themselves of power plants. The new power plant owners would not be under the obligation to sell into the Power Exchange and could sell power to a direct access customer if a better price could be obtained.

Since there is a significant surplus of power generation in the western United States, the competition created by the Power Exchange would initially be based on the short-run marginal costs of existing power plants, which are quite low. Over time, market participants would gain experience in determining which generation units cleared the market price—and were therefore dispatched—and how much income they produced. Non-utility generators would have access to this information and would be able to determine whether it was in their interest to bid into the Power Exchange or to seek direct access customers. As the supply of wholesale power came into a better balance with demand, the price signals sent by the Power Exchange would help inform decisions about building new power plants. An investor or developer would know what percentage of the time the spot price exceeded the capital and operating costs of a new power plant and could then calculate the risk of earning a reasonable return on such investments.

Direct Access

Under pressure from industrial customers, the CPUC decided to allow for phasing in of direct access transactions between end users of electricity and suppliers, including brokers and marketers of energy services. The CPUC agreed to a five-year phase-in period starting in 1998. Under the proposed schedule, all customers would be allowed to choose a provider by 2003. However, the CPUC cautioned that there might be problems with this approach that would need to be evaluated along the way. The phase-in schedule would allow the CPUC and other stakeholders to determine how bilateral contracts affected the operation and management of the transmission system.

Consumer groups recognized that there would be significant barriers keeping small business and residential customers from effectively participating in the direct access program. They argued that it was necessary to aggregate their relatively small loads so that cost-effective transactions based on economies of scale could be carried out. The CPUC agreed that power suppliers, as well as third-party intermediaries, would be allowed to aggregate customer loads for the purpose of marketing electricity. These companies would be allowed to offer electricity and other energy services, such as DSM programs, to meet specific customer needs.

The Independent System Operator

In order for any deregulation model to work, transmission needs to be available to all generators of power on a nondiscriminatory basis. A number of approaches have been proposed for reforming the transmission grid, including forcing utilities to divest their ownership of high-voltage power lines as well as schemes that allow for continued ownership but without day-to-day operational control. In 1992 Congress passed the Energy Policy Act, which called for nondiscriminatory open access to all buyers and sellers wishing to rely upon existing transmission paths. In the past, utilities have offered transmission service access on terms that raised the price for alternative supplies and that in turn made their power generation the lowest cost option.

A key to making competition work is to eliminate favoritism in decisions about getting electricity from where it is generated to the switchyards and substations from which local distribution utilities then deliver it to customers. Under the CPUC proposal, the state's three private electric utilities are to turn over control of their transmission facilities to the ISO to lessen the opportunity to influence decisions that favor their own power plants over those owned by competitors.

Utility Distribution Companies

The remnants of PG&E, SCE, and SDG&E will continue to own and be responsible for the operation and maintenance of the substations and wires that connect customers to the transmission system. These utility distribution companies will have an obligation to provide distribution service to all customers in their service area, including direct access customers. The rates they charge for this monopoly service will be subject to CPUC oversight. The utility distribution companies will also have an obligation to serve all customers who do not elect to take service from a direct access provider. They can purchase all or a portion of the power needed to serve these customers from the Power Exchange or from other sources. PG&E, SCE, and SDG&E would be prohibited from entering into contracts to purchase power from facilities in which they have an ownership interest either directly or through an affiliated company.

The CPUC indicated that it wanted to use performance-based ratemaking (PBR) to encourage utilities to provide this service efficiently. How well the utility distribution company performs using these measures will determine its compensation as a regulated natural monopoly.

In traditional rate-of-return regulation utility managers were protected from most risks. Rate-making mechanisms allowed utilities to pass through all costs associated with purchased power to its ratepayers. Risks associated with weather, forecasting, and demand-side management were also eliminated in California, a by-product of well-intentioned efforts to sever utility incentives to sell, instead of conserve, electricity. PBR has been heralded as a device designed to reward superior performance. Whereas traditional regulation determined whether expenditures were prudent after the fact, PBR rewards performance measured against specific benchmarks.

The simplest form of PBR is a price cap in which the utility is given a maximum price that it can charge for a service. The theory is that this type of arrangement gives utilities an incentive to keep costs low, since every dollar saved goes to the utility's bottom line. In practice, such schemes are less reliable. Price caps are often altered in response to costs incurred by the regulated entity. Regulators are unlikely to stick to a price cap that threatens the financial health of the utility.

A modified price cap approach—in which the regulated entity gets to keep some percentage of its cost savings or is required to pay for some portion of the amount by which it exceeds the cost standard—can work better than a pure cap in that it makes it easier for the regulator to maintain set formulas for performance. This has been the approach favored by the CPUC. Still, the basis for determining the benchmarks is still problematic.

The PBR proposals put forward by California utilities have raised alarm. David Morse, an analyst with the CPUC's Division of Ratepayer Advocates (DRA) took a close look at the initial PBR proposals put forward by the state's three investor-owned utilities in the early 1990s and didn't like what he saw. Though SDG&E had some new risks incorporated into its PBR, both of California's largest utilities continued to insulate managers from risk, claimed Morse. For example, both PG&E and SCE would still be able to pass through to ratepayers 100 percent of all variable costs.

The CPUC's December 1995 restructuring proposal calls for performance measures to be established for reliability, safety, and quality of service.

Stranded Costs

The policy recommended by the CPUC for stranded costs is to allow them to be collected in a manner that is competitively neutral, fair to all classes of customer, and does not raise rates. This last consideration was an admission that utility reform in California would not reduce rates, at least not in the short term. Many of the proponents of competition and the restructuring of electric utilities, particularly organizations representing large industrial customers, were disappointed by this concession.

The CPUC had proposed to compensate the owners of stranded assets by imposing a charge that could not be bypassed, called the Competition Transition Charge (CTC), on all current retail customers of the state's investor-owned utilities, regardless of who they purchased power from in the future. They also proposed that rates would be capped at their January 1996 level for all customers who continue to purchase electricity through existing utilities.

The decision to allow utilities to recover all of their stranded costs was explained as necessary to assure their financial integrity. The CPUC was also concerned that utilities not be hobbled participants in the future competitive market. Its proposal guaranteed that utilities would be able to completely recover all costs associated with contracts for power; that previous commitments made for nuclear power plants would be honored; and that the cost recovery of other stranded utility assets would be accelerated. The profits earned on the assets slated for accelerated recovery would be reduced by 10 percent to reflect the reduced utility risk. However, the utility could make up that loss in profits by divesting themselves of their power plants. Costs associated with retraining and early retirement of utility employees affected by restructuring could also be recovered through the CTC.

Since no existing costs would be eliminated during the transition period, and new costs could be added, some questioned how California's rates would go down. As previously noted, some stranded costs are self-correcting. This is the case with independent energy projects. If the CPUC had done nothing, rates would have gone down in the future because of the "Year 11" provisions included in existing power purchase contracts. The promise by the CPUC to cap rates actually meant that the state's investor-owned utilities would increase their revenues during the transition period to pay off investments in uneconomic power plants.

The transition to a fully competitive market in California would last through 2003. All stranded costs—except those associated with longer term contracts—would have to be collected by 2005. After that, the CPUC envisioned that competition between electricity suppliers would really take off.

The details of setting the CTC will keep armies of accountants busy for years. The CPUC outlined some of the details in its restructuring proposal, but many were left for the implementation phase. The CPUC recognized that a competitive market for electricity will categorize power plants as either economic or uneconomic, in whole or in part. Some plants might be economic for some periods of the year and uneconomic in others. In the simplest terms, a power plant is uneconomic if its net book value (that is, its original cost minus depreciation) is greater than its market value. For each utility, the transition costs are the combined above-market costs of all its assets, both economic and uneconomic. One estimate of each of the three utility's generation stranded costs is presented in Figure 4-5. Also included is a net present value estimate of California's QF capacity by 1999.

Transition costs will be quantified in two ways. For those assets that are sold, a comparison of the book value and market value will be relatively straightforward. For uneconomic plants that continue to operate during the transition period, these transition costs will be the difference between the market price and the cost of producing the power, minus whatever residual value the plant has after the transition period. This is the area in which calculation of costs gets sticky. Transition costs can arise from the continued operation of power plants, particularly nuclear facilities that might have to be modified for safety reasons. For that reason, the CPUC dealt with each of the nuclear facilities owned by the state's utilities in special ways.

The SONGS Settlement

The question of how to treat nuclear stranded costs first arose in a debate about the two operating nuclear power plants at the San Onofre Nuclear

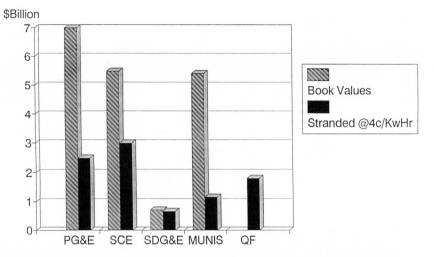

Figure 4-5: This bar chart displays the levels of stranded generation investment for California's investor-owned utilities as well as the aggregated estimates for all of the state's municipal utilities. Called out separately are costs associated with QF independent power projects as of the year 1999. *Source:* Center for Energy Efficiency and Renewable Technologies.

Generating Station (SONGS). SONGS is located on the Pacific Coast between the residential communities of Oceanside and San Clemente, California. SONGS originally comprised three nuclear reactors but the oldest one was closed in 1992 after SCE admitted that it was no longer economic to operate.

SCE and SDG&E (which has a 20 percent ownership share of the remaining two reactors) requested that the CPUC allow an accelerated recovery of their sunk capital costs. A settlement was reached in 1994. The proposed SONGS settlement would allow the utilities to recoup their unamortized investment of $2.7 billion in eight years instead of twelve. In exchange for this accelerated recovery, the rate of return on this investment would be reduced from 9.8 percent to 7.78 percent. Total power costs—including the accelerated capital cost recovery—would keep prices at the current level (approximately 8 cents/kWh) until 2004. At that point, the utilities could sell power on the open market.

Before the SONGS settlement was approved, however, several consumer groups intervened, arguing that the settlement allowed SCE to increase rates above what they had previously agreed to for the next eight years. SCE claimed that by lowering its rate of return on the nuclear facilities, ratepayers would save about $1 billion compared to traditional ratemaking. Representatives of industrial customers complained that ratepayers would not see any of those savings until 2003.

The Coalition of Utility Employees, which represented 40,000 utility employees, also objected to the settlement. Their attorney complained that the proposed agreement would reward SCE whether or not the facility continues to operate. He thought that the settlement provided an incentive to close the reactor, which would result in significant employee layoffs.

Environmentalists were closely following a separate regulatory proceeding in which SCE had proposed scaling back a marine mitigation plan designed to help reverse the toll the reactor had on the surrounding coastal environment. A 15-year study had found that the plant's cooling system takes in as much as 58 tons of fish every year. Four billion eggs and larvae were also part of the reactor rinse.

In late 1995 the California Coastal Commission (CCC) denied the utility's request to scale back the measures designed to compensate for the nuclear plant's environmental impacts. Before the vote, an SCE representative told the *Los Angeles Times* that moving forward with the more expensive measures might make the plant unprofitable. Environmentalists discovered that SCE submitted different estimates of the costs of the mitigation plan to the CCC and the CPUC. The controversy revolved around the costs of marine environmental mitigation contained in a PBR scheme. The purpose of the PBR mechanism was to create incentives for utility shareholders to lower the cost of electricity from SONGS that was sold to consumers.

SCE had proposed to capture the costs of its marine mitigation plan through its rates. The utility told the CPUC that it would need to collect $80 million to pay for its plan. But it was also seeking relief from the CCC to lower the cost to $53 million. Some observers argued that SCE wanted to pocket the difference. Whether or not this was true, the mitigation issue raises several questions about policies for restructuring electric utilities. Such PBR mechanisms might provide incentives for utilities to cut corners on implementing environmental or other programs that provide broad benefits, such as public safety. Another issue relates to the definition of stranded costs. The marine mitigation measures were future costs that SCE is required to expend to meet regulatory commitments. In a competitive market, any future environmental costs are simply part of the risk of doing business. If environmental problems force a power plant's owner to make additional investments, the owner has to decide whether they can be recouped in the market. Otherwise, the facility will be closed because it is no longer economic. The SONGS settlement would allow SCE to recoup future investments, which would further raise the costs of the already expensive power generated at the nuclear plant.

Opponents of the settlement argued that it should be incorporated into the broader restructuring decision that the CPUC was considering. Though CPUC President Dan Fessler did not support those petitions, he did raise an interesting question about the settlement. He noted that SCE ratepayers would pay $576 million more for SONGS electricity over the next eight years than they would otherwise. And after SONGS was fully paid off in 2003, its owners would be free to sell power in an open market. If SONGS was competitive beyond 2003, the ratepayers who had paid for the reactor might not receive any future benefits from it. Fessler proposed a 50-50 split of all future benefits related to the operation of the San Onofre plants. This proposal by Fessler raised the question of whether accelerating the cost recovery from existing power plants was granting utilities an unfair competitive position in the future.

The CPUC finally approved the settlement at the beginning of 1996. It reduced the rate of return on the investment in the plant to 7.8 percent on debt and 7.4 percent on equity. These figures would be similar to the general policy that the CPUC was proposing for treating generation assets under restructuring, allowing utilities to earn 90 percent of their expected profits on stranded assets. If SONGS were closed before 2003, the remaining capital investment would be at risk. Furthermore, SCE ratepayers were to receive 50 percent of any benefits from SONGS operation after the year 2003 (though it was unclear how such "benefits" would be defined or awarded to ratepayers).

Environmentalists and consumer groups were concerned that the SONGS settlement would become the cost recovery model for uneconomic nuclear power plants all over the country. Their principal concern was the decision to allow future costs, such as marine mitigation or safety upgrades, to be defined as stranded costs. This treatment could allow nuclear plants to continue running beyond their useful economic life and impede the introduction of newer, more efficient technologies. Utility investors, in general, approved of the SONGS settlement. The CPUC had allowed investments in the two nuclear power plants to be recovered on an accelerated basis and had protected, for a while, their continued operation from the discipline of the market.

Market Power

Many participants in the restructuring debate were concerned that the state's big three investor-owned utilities would exploit market power in an unregulated market setting. The utilities might have the potential to exercise market power in an unregulated wholesale power market because they would control most of the power plants in the state. Both SCE and PG&E

currently meet more than three-quarters of the demand for power in their respective markets; SDG&E meets 70 percent of its demand. These figures underscore how these investor-owned utilities could dominate a future market if they continued to own most of the power generation in the state.

The Federal Trade Commission informed the CPUC that PG&E and SCE owned sufficient generating units that they were likely to provide the final increment of power needed to meet the demand for electricity for many hours of the year. By controlling this increment of supply needed to clear the market, the utilities could manipulate the price to be set by the Power Exchange. The CPUC ordered PG&E and SCE to submit plans to divest themselves of at least 50 percent of their fossil fuel–powered generating assets. They also agreed to provide incentives to encourage further divestiture.

Another concern of consumer groups was the potential for cross-subsidization between the regulated and unregulated parts of electric utilities. Under the CPUC restructuring proposal, the distribution utility would still be regulated, while the utility generation company would not. The potential exists for utilities to shift costs from their competitive services to their regulated services. This both gives the utility an unfair advantage in the marketplace and increases the cost of the regulated monopoly service.

Finally, consumer groups were worried about the potential for predatory pricing by utilities to drive competitors out of the market and to erect barriers for new market players. *Predatory pricing* is the practice of selling a good below costs in order to undermine competition. The technique was employed by America's railroads at the turn of the century. It is now illegal under the Clayton Anti-Trust Act of 1914. In order for predatory pricing to work, a company must be able to withstand the short-term losses while absorbing the increased demand. The firm must also be able to secure future profits through monopoly pricing. To sustain a monopoly situation, there must be significant barriers to entry in the market, such as high initial investment costs. A number of observers believe that an unregulated electricity market in California could create conditions leading to predatory pricing.

The CPUC believes that the creation of an ISO, when coupled with a requirement for divestiture and the separation of transmission, distribution, and generation services into three separate subsidiaries, will be sufficient to prevent predatory pricing. Yet it also acknowledged that behavior would have to be closely monitored in a dynamic and evolving competitive market.

Stranded Benefits: Energy Efficiency

Under the existing California regulatory framework, utilities are allowed to carry out DSM programs that meet cost-effectiveness tests. State law supports utility involvement in delivering energy efficiency measures that are not being carried out by other market players. As discussed earlier, the primary motive for the state's encouragement of investment in energy efficiency has been the desire to defer or avoid building or using more expensive and higher polluting electric generation.

Program's carried out by California's investor-owned utilities have been some of the most successful in the nation. The CEC claims that with the utility investments in energy efficiency measures from 1990 through 1994, California will save ratepayers $2 billion over their useful life (see Figure 4-6).

Supporters of utility energy efficiency initiatives were worried that these programs might suffer as competition was introduced into California's electricity market. Power generators would, once again, see an economic benefit in promoting power sales. And utilities would cut back on DSM programs to focus on short-term costs so as not to lose customers, particularly large industries, to competing suppliers. These concerns proved to be more than hypothetical.

After the CPUC released its first restructuring proposal in April 1994, all three of the state's investor-owned utilities reacted by gutting energy effi-

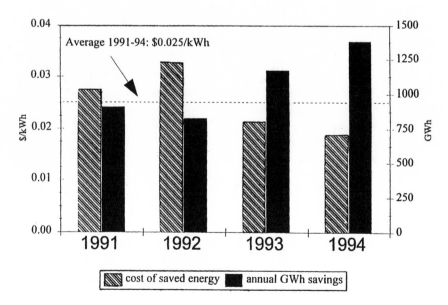

Figure 4-6: These figures show that energy savings derived from California's investor-owned utility DSM programs cost an average of 2.5 cents/kWh between 1991 and 1994. *Source:* Peter Miller, NRDC.

ciency programs. Expenditures on these programs dropped by 30 percent between 1994 and 1995.[7] The decision of the state's three investor-owned utilities to slash these programs raises an important question. How big an impact did California's energy efficiency programs actually have on utility rates? NRDC undertook an assessment of energy efficiency and other public benefit program expenditures made by California's three investor-owned electric utility expenditures in 1993 and 1994. They found that they totaled 3.4 percent of electric utility revenues in 1993 and 1994.

Although these programs were minuscule in comparison with other utility expenditures, they were the first to be cut by panicky utility managers worried about the effects of the CPUC restructuring proposal. NRDC estimates that over the next 20 years SCE ratepayers will pay $750 million more for electricity if the cuts made to demand-side management programs are sustained in the future. The reductions have also harmed the kinds of pioneer companies California has been trying to attract for new economic development.

One company hurt by the CPUC restructuring proposal was Appliance Recycling Centers of America (ARCA), a Minnesota firm that had invested $5 million in a refrigerator recycling facility in Compton, California. The facility not only recycled all the chlorofluorocarbons (CFCs) in the refrigerant (commonly known as freon) but also captured all the CFCs contained in the foam insulation used in the refrigerators' walls and doors. The release of these chemicals into the atmosphere has been identified as the major cause of the depletion of stratospheric ozone, which protects humans from the harmful effects of ultraviolet radiation.

The program to recycle refrigerators had been hailed by SCE as part of its commitment to "Rebuild LA," an effort to revitalize regions impacted by the 1993 riots. SCE provided rebates to customers to replace old inefficient refrigerators with more efficient models. The old refrigerators were picked up and turned over to ARCA for recycling. The program resulted in saved energy, lower electric bills for consumers, new jobs for residents of the inner city, and an improved environment.

But when SCE proposed cuts in its conservation spending, it lowered its previous commitment to supply ARCA with 177,000 refrigerators over a three-year period. SCE originally projected that its refrigerator replacement program would net 33.4 megawatts of capacity savings. However, with a reduced energy efficiency budget, the utility reduced by more than half the business ARCA was counting on. The company was forced to cut back on its operations and reduce the size of its work force.

The rationale for restructuring the electric services industry is to create a better business climate. Yet the focus of utilities on cutting their energy efficiency budgets has damaged companies like ARCA, a national leader in utility recycling programs. Throughout the United States, 54 million major household appliances will need to be managed through disposal or recycling by the year 2000. If those appliances are not recycled, they will likely eventually end up in landfills, releasing their CFCs into the atmosphere. Though SCE later reversed course and promised ARCA it would restore the program to its original goals, the uncertainty and delays set back the company's plans for expansion of its business in California and elsewhere.

V. John White, executive director of the Center for Energy Efficiency and Renewable Technologies (CEERT), stated that the cuts in the utilities' energy efficiency budgets "showed the utility priorities" under restructuring. "Utilities such as Edison want to cut the programs that provide the most value to residential customers," said White. He cited a 1995 poll financed by environmental groups that showed that Californians opposed trade-offs between nuclear and conservation/low-income programs by a margin of 74 percent to 16 percent.

DSM programs have proven to be an effective low-cost resource for the utilities. By 1994 the average cost of these programs was approximately 1.9 cents/kWh saved, and the total costs to ratepayers were very small. The CEC estimated that the average residential customer in the state paid about $14 a year to support energy efficiency programs.

To prevent a further deterioration in California's energy efficiency market, policymakers and energy efficiency advocates began to look at mechanisms that would preserve the benefits of integrated resource planning and DSM programs in a manner that was compatible with more competitive electricity markets. The CEC identified three steps that could be taken to maintain or increase the benefits created by utility DSM programs. The first is to separate the generation and distribution portions of utilities to eliminate financial conflicts of interest between the sale of energy as a commodity and energy efficiency as a service. The second is to use PBR mechanisms that link revenue recovery to the number of customers served by the distribution utility rather than to the number of kilowatt hours sold. The third is to create a nonprofit organization to oversee the use of ratepayer-funded expenditures on energy efficiency measures.

The CEC urged that a "two-track approach" be adopted to stimulate the continued development of a DSM market in California. The first track would be for utilities and competing firms to offer customer-value DSM measures in a competitive market. The second track would be to fund cost-effective energy efficiency measures that would not be supported in a competitive market through a surcharge on all consumers. Restructuring would require that incentives for energy efficiency measures be different than the ones used for vertically integrated utilities.

The CPUC embraced the two-track approach for energy efficiency programs. Market-driven funding mechanisms would support most DSM projects targeted to individual customers (track 1). A surcharge would be levied on all customers to support activities to transform energy efficiency markets and to educate customers about effective conservation measures (track 2).

A coalition of utilities, private energy service companies, federal and state agencies, environmentalists, and installation contractors recommended that an equitable charge that could not be bypassed be put in place to support energy efficiency programs through a transition of at least five years.

The CPUC recognized broad support for a "public benefit charge" for funding energy efficiency and load management programs. However, they noted that there was not agreement on how to draw the line between the two tracks. They recommended that funds collected through the public benefit charge should eventually be administered through an independent, nonprofit organization. However, during a transition period the CPUC advocated tapping into the expertise that the state's utilities had developed in administering DSM programs.

Renewable Resources

California has one of the most diverse mixes of supply-side resources of all states in the country. The CPUC indicated that it wanted to maintain this diversity as well as encourage the development of new renewable resources. Several ideas were submitted to the CPUC for achieving these goals. The Environmental Defense Fund suggested that renewable resources should be selected through a competitive bidding process. The funds that would be needed to pay for the higher costs of the resources would be obtained through a surcharge on all retail electricity sales, as was being recommended for energy efficiency services.

However, after their experience with the BRPU, the CPUC was reluctant to become involved in another auction to select new power sources. Instead, it recommended that a minimum renewable purchase requirement be adopted to meet the state's goals for maintaining and increasing diversity of energy supplies. The requirement was based on a proposal put forward by the American Wind Energy Association (AWEA) called a "renewables portfolio standard."

The objective of this policy was to ensure a minimum level of renewable energy in the California resource mix in a way that promoted competition within the renewables industry. To do that it would be necessary to create a predictable market for renewable energy sources that enabled financing at reasonable costs, minimized barrier to entry into the market, and kept transaction costs low. It was also important that the policy achieve competitive neutrality, that is, it would apply equally to all competitors in the restructured marketplace. Finally, to achieve public support it had to avoid windfall profits or unexpected impacts on rates.

As envisioned by Nancy Rader, a consultant for AWEA, the renewables portfolio standard would work as follows: As a condition of doing business, power suppliers including the utility distribution companies, direct access suppliers, and power aggregators would be required to demonstrate that a percentage of the energy they sell to end users comes from renewable resources. The percentage would be determined by the state (for California the 1994 level of electricity coming from non-hydro renewables is 11 percent) and would increase over time. The selling of renewable kilowatt hours to retail customers or the Power Exchange would result in the creation of a "renewable energy credit" that would be tradable. Individual energy suppliers would not need to develop renewable energy sources or even purchase power from a renewable power plant. They could meet the requirement by purchasing renewable energy credits from other suppliers of renewable energy. This market-based approach would allow the standard to be met in a cost-effective way. Government involvement would be limited to monitoring and enforcing compliance, as well as setting the initial market rules.

The idea is simple and attractive. It is modeled on the approach used in the Clean Air Act Amendments of 1990 for reducing national emissions of sulfur dioxide. However, there are a number of important details that need to be resolved. Its successful implementation will require that investors in renewable power plants have confidence that the policy will remain in place long enough to allow them to recoup their investments. At a political level, it will also have to account for initial differences in the competitiveness of renewable technologies. In California, wind and geothermal technologies are currently the lower cost renewables. Biomass and solar thermal technologies are more expensive, and photovoltaics are the most expensive. Finally, details for certifying the credits and monitoring their purchase need to be worked out, as do compliance mechanisms and penalties for noncompliance. An attractive feature of the renewable portfolio standard is that it is compatible with the current structure of vertically integrated monopolies and wholesale competition or with a restructured industry that relies upon distribution-only utilities to deliver electricity to retail customers. For more details about implementation issues surrounding the renewable portfolio standard, see Appendix B.

Aftermath

Making public policy is often compared to making sausage. It is messy and the faint-hearted are advised not to look too closely. The two and one-half years the CPUC spent revising its various restructuring proposals was a haphazard journey that bordered on chaos. Certainly, there was much arrogance in the CPUC's belief that such a radical proposal as "retail wheeling" could be imposed upon the electricity consumers of California with little public discussion. Looking back at the original 1994 proposal, it is remarkable that CPUC President Dan Fessler expected it to have the force of state law just four months after being presented to the public.

In retrospect, it is clear that the CPUC erred in not involving the public early on in a discussion about the need for restructuring and the goals the public wanted to achieve in a new electric order. Their approach of unveiling a detailed proposal through a simple press release forced public interest groups to take an adversarial position. The turnout at public hearings was large and angry. Two of the commissioners (one a former Congressman) who had supported "retail wheeling" decided to leave the CPUC rather than fight for their point of view. California's governor, who had been passive on the issue of restructuring, became disturbed when two of his important constituencies, large industries and large utilities, ended up on opposing sides of the first two proposals. His intervention led to the compromise pool/direct access proposal now being discussed in California.

Given the contentiousness of the California process, it is remarkable how much all of the participants—utilities, consumer groups, and environmentalists—have learned. Entrenched positions changed as the debate was joined. The public hearing process was an eye-opening experience for the investor-owned utilities. According to Kathy Treleven, a strategic planner at PG&E, the free-wheeling discussions about restructuring issues was odd. "We are used to operating in a judicial environment. We make our case and so does the opposition and a judge splits the difference. This process was more like surfing," said Treleven, noting that the process seemed to meander as political forces coalesced in different ways. Operating in a restructured world will also be a new experience for utilities. While changes might be "invisible" to customers, PG&E will have to make dramatic internal changes. The creation of a wholesale power exchange and an independent transmission system operator will transform the way utilities make decisions about investments in power plants. "We now have to bid . . . to run our own units," Treleven observed.

One major insight culled from the California experience is that a process to restructure the electric service industry should be bottom-up, rather than top-down. Critics of the CPUC process, such as Jan Hamrin, founder and former director of the Independent Energy Producers, has harsh words for California's haphazard approach. "Other states have learned from California and are relying much more on a collaborative model," she stated. However, Hamrin thinks the CPUC ended up with a reasonably good market structure. Still, she believes that annual adjustments in the CTC to cover stranded costs gives utilities an incentive to manipulate the wholesale power market and shut out new technologies. Nonetheless, the CPUC's adoption of a renewable portfolio standard and systems benefit charge were features that have wide application elsewhere. Hamrin's key advice: deal with stranded costs early on "because it drives everything else."

For environmentalists such as NRDC's Ralph Cavanagh, who has worked tirelessly within the existing utility monopoly structure to create incentives for energy efficiency programs, the debate over restructuring in California stimulated a major rethinking of the role of utilities as agents of social policy. Cavanagh was among those leading the charge against retail wheeling after the CPUC unveiled its first restructuring proposal. From his view, the biggest mistake in California was the damage done during the transition to a more competitive marketplace. Other states should implement a systems benefits charge to maintain momentum in energy efficiency programs while debating restructuring, he recommends. Washington Water & Power and PG&E have now instituted such charges to recover DSM investments. According to Cavanagh, use of such charges can, if properly designed, avoid jurisdictional disputes between the states and FERC.

Cavanagh warns that "formidable barriers continue to obstruct efficiency improvements that are highly cost-effective by almost any measure." Restructuring is reinforcing these barriers in the short term. One impact of a more competitive market, notes Cavanagh, is an identified need for less costly administrative mechanisms to certify and reward performance in the delivery of energy efficiency services.

Though Cavanagh continues to believe that utilities remain important vehicles for promoting DSM, he recognizes that "there is no natural monopoly involved in the actual delivery of energy efficiency improvements." Utilities were convenient agents due to their captive customers and resources. The ultimate solution, says Cavanagh, is disaggregation of the vertically integrated utility monopoly. Despite the progress NRDC has made in prompting investor-owned utilities to be responsible corporate citizens, Cavanagh is now convinced that the monopoly structure can no longer hold because of "increasingly obvious embedded conflicts of interest." He believes that the responsibility for power generation and distribution need to be completely separated so that the company that owns the wires connected to customers' homes and businesses has no interest whatsoever in who produces the electricity.

Cavanagh thinks power-generating companies will sell electricity into a deregulated commodity marketplace. Distribution utilities, on the other hand, should be designed to minimize customer bills through DSM and other techniques. His remarks reflect an evolution in thinking among one of the leading environmental organizations involved in energy issues, the NRDC. Its possible that an even bigger leap in thinking was taken by advocates of consumer interests. Lenny Goldberg, a lobbyist for residential ratepayers, now compares the restructuring of vertically integrated utilities to the breakup of the Soviet Union. Utility monopolies are analogous, in his view, to the central planning paradigm that was dismantled with the dissolution of the old Soviet empire. "The rise in democracy can be equated with competition," Goldberg states. "The authoritarian monopolies are trying to stall the integration of new democratic markets. All over the country, utilities, like the vestiges of the old Soviet Union communist system, are using their political power to protect market share."

Goldberg works on behalf of Toward Utility Rate Normalization (TURN), a San Francisco–based public interest group that has helped create programs such as lifeline rates and low-income customer assistance under the old regulatory framework. For TURN to endorse moves to a more open marketplace for electricity is a major shift in thinking among those who have viewed deregulation schemes in other industries with much skepticism.

Growing numbers of environmentalists have joined TURN in viewing the restructuring of the electric service industry as an opportunity to allow citizens to have a greater say in where their power comes from. Among them is V. John White, executive director of CEERT, a Sacramento-based coalition of environmental groups and private sector energy services and renewable energy development companies. "Deregulation is both a crisis and an opportunity for environmentalists. It could be good or bad for the environment. It all depends upon how it is done," remarked White at a Sacramento conference in January 1996. He went on to add that although environmentalists "lose our traditional regulatory levers," under a deregulated system, price competition could also force early closure of uneconomic and environmentally damaging sources of electricity, such a nuclear power plants. "What we will have to rely on is the marketing of green power to people to reduce air pollution from the electric utility sector," White said.

Nonetheless, White warned that a danger associated with restructuring is that "the clean power guys get shoved outside of the system." The fact is that during the past two and one-half years in which restructuring has been debated, many have been financially damaged. "The real risk is what happens between now and when we get to the future model of the electricity marketplace," he noted.

White and environmental groups such as the Sierra Club and the Environmental Defense Fund decided that California's system of integrated resource planning "was no longer sustainable" in light of political and economic trends. In California, White points out, the advantages of competition did not register with the environmental community until late in the debate over restructuring. That allowed utilities and large industries to cut a deal—the MOU—while consumer groups and environmentalists were on the sidelines. White warns that if this occurs in any other state contemplating restructuring, "environmental values will be in jeopardy."

What should be the ultimate goal of restructuring? Environmentalists say the goal should be new investment in cleaner sources of energy. If this becomes the objective, then an alliance could be forged among small and large customers, environmentalists, and new market entrants. New investment would allow these major participants to create a common strategy to phase out older, inefficient, and dirty power plants.

Despite the failure of the California restructuring process to focus on a strategy to phase out uneconomic power plants, much was still accomplished. There is general agreement in support of a universal system benefit charge (that cannot be bypassed) for energy efficiency, low income, and R&D programs. It is agreed that transition costs will be fairly allocated among all customer classes. Both an independent transmission system operator that will have no economic interest in power plants and a voluntary statewide wholesale power pool called the Power Exchange will be created.

Notes

1. The largest single source of "stranded" costs nationwide is the $70 billion in nuclear capital investment—a figure that represents 60 percent of the book value of America's 108 operating reactors, according to a report issued by RCG/Hagler Bailey in 1995. The calculation assumes full-scale competition occurring throughout the country by 1999. SCE was ranked as having the third highest exposure of utilities in the country (behind Texas Utilities and PECO.) A ranking of nuclear facilities in terms of stranded cost totals placed Palo Verde as third and SONGS as eighth highest in the nation. (The two highest ranked facilities—Comanche Peak and South Texas Plant—are both located in Texas.)

2. Black claims that the $250 billion labeled "stranded costs" is a measurement of inefficiency. To put this large number in perspective, he estimates that with absorption of these costs by the marketplace "utility bills would drop about 15 percent on average— far more in some geographic areas; not at all in others." If utilities are allowed to recoup 100 percent of these costs, the largest financial rewards would go to those utilities that are the least competent competitors. According to Resource Data International (RDI), over 50 percent of the nation's stranded cost liability is concentrated in a handful of states: California ($16 billion); Illinois ($10 billion); New York ($14.3 billion); Ohio ($11.5 billion); and Pennsylvania ($12.6 billion).

3. For example, CLECA members employ 3,500 people. CLECA's 11 members require 330 MW of electricity for 14 different facilities. CLECA's jobs-to-electricity ratio is less than 10 jobs per MW. Given that the state employs over 14 million people, the statewide average is more than 300 jobs per megawatt.

4. A report released in August 1995 by Resource Decisions for the California Energy

Commission entitled "The Effects of California Electricity Market Restructuring on Emerging Technologies," offers evidence that the POOLCO model would limit opportunities for renewables. According to Richard McCann, "POOLCO eliminates niche markets" by maintaining a single price for electricity. Co-author Marvin Feldman added that direct access would "allow green energy providers to market to consumers willing to pay more for more environmentally benign megawatts."

5. "POOLCO May Become MOU as SCE, Industrials, IPPs Negotiate," *IRP Report* (New York: McGraw-Hill), September 1995, p. 15.

6. The parties that signed on to the Framework principles include Utilities Consumer Action Network, Union of Concerned Scientists, Toward Utility Rate Normalization, Sierra Club California, Public Citizen, Natural Resources Defense Council, Environmental Defense Fund, Center for Energy Efficiency and Renewable Technologies, California Public Interest Group, California/Nevada Community Action, and American Wind Energy Association.

7. "California Coalition Calls for a Separate Fee to Fund DSM in Deregulated Arena," *Demand-Side Report* (New York: McGraw-Hill), October 12, 1995, p. 4.

Chapter 5
Regional Examples of Utility Reform

"Our search for reliable information is itself guided by the
questions that arise during arguments about a given course of
action."

—Christopher Lasch

The states have been called the laboratories of democracy. Although the
laws of physics governing electricity are the same for all 50 states, the laws
put in place by policymakers to regulate electric services vary widely. Energy
policies are shaped by cultural values and political institutions that are
unique to particular parts of the country. Despite this diversity, ideas and
strategies often cross state lines. State legislatures and regulatory bodies
naturally look to what is happening elsewhere to see what might work in
their area. The demand for an overhaul of the existing regulatory maze governing electric utilities has spread across the country, with many PUCs now
considering how to tap into the forces of competition to lower the cost of
electricity.

Though circumstances motivating moves toward competition differ, there
are common threads that tie these regional stories together. Interestingly,
the areas in which the debate on restructuring is the most intense are those
where past nuclear investments created a financial crisis. As in Sacramento,
a troubled nuclear program often prompts a community learning process. At
first, utilities usually resist more citizen involvement in utility decision making. Yet citizen groups have found ways to take the initiative and offer alternative energy strategies that can accomplish more with less. Often, these ef-

forts by the citizenry have instilled more economic discipline in utility planning.

Cultural and historical differences in the development of the economies of New England, the Pacific Northwest, the Tennessee Valley, Texas, and Minnesota played an important role in forming the current electricity delivery systems in these areas. They may also influence the design of new systems. In each case, the spark for change has come from groups of citizens. Often, environmental organizations would lead the way in infusing greater accountability and competition into resource planning.

New England

One week to the day after Rancho Seco was closed by the voters of Sacramento, the Seabrook nuclear power plant in New Hampshire went critical. At 5:23 p.m. on June 13, 1989, operators began a sustained nuclear fission reaction in the plant's 100 tons of uranium fuel. That event punctuated a decade-long regional debate about energy policy that involved state governors, U.S. senators, and thousands of ordinary citizens.

The New England electricity market is very different from California's. The region consists of six small states, each with its own PUC. The area is served principally by investor-owned utilities, many of which operate in more than one state. The principal source of renewable energy that has been developed in the region is biomass, although one of the world's largest wind turbines was built in Vermont in 1941. The region has proven to be fertile ground for the development of innovative energy conservation programs. Environmental advocates first became involved in energy policy debates as a result of the failure of the large nuclear program at Seabrook, New Hampshire.

Seabrook is the site at which several utilities decided to build two nuclear reactors. One was abandoned in 1982, only 25 percent complete. The other eventually cost $6 billion and bankrupted its lead owner, Public Service of New Hampshire (PSNH). The 11 other New England utilities owning plant shares were also financially stressed by the nuclear program. This nuclear dilemma spawned efforts by citizen groups to reform resource planning in the region.

New England, unlike California, actually experienced a shortage of power supplies in the mid-1980s. Delays in the start-up of Seabrook and rapid economic growth had pushed the region's energy supplies to the limit. Worried that industry might relocate because of blackouts and brownouts, the New England Governors' Conference undertook a study in 1985. The resulting report predicted demand for electricity in the region would likely grow by

2.2 percent annually until the year 2000. It called for quickly developing new power supplies.

The Conservation Law Foundation (CLF), a Boston-based environmental group, was concerned that the governors' plan would rely primarily on the construction of new power plants and large power purchases from Canada to meet the region's growing demand for electricity. They argued that "least cost" or integrated resource planning (IRP) should also look at the potential for reducing power demand through energy efficiency measures.

By the mid-1980s, CLF had built a reputation as a tough litigator for environmental causes. CLF stopped offshore oil drilling at Georgia's Bank, Massachusetts—one of the nation's most productive fisheries—by making the case that damage to the fishing industry exceeded the benefits that might be derived from petroleum sales. The next case in which they were to hone their skills in using persuasive economic arguments was Seabrook 2. CLF had avoided being drawn into the controversy around Seabrook 1. Opposition to the first reactor was led by the Clamshell Alliance—a diffuse organization that favored civil disobedience and symbolic direct action as practiced by some in the civil rights movement. CLF was convinced that they could persuade utilities investing in the second reactor that there was a less costly way to meet the region's energy needs. They would argue that this alternative approach would also be less risky to shareholders.

CLF and the Union of Concerned Scientists brought in a number of experts to look at alternatives to Seabrook 2, including greater reliance on energy efficiency measures. CLF decided not to challenge any of the assumptions underlying the cost projections for Seabrook 2, even though they knew they were very questionable. Instead they demonstrated that reasonable alternatives based on conventional technologies would cost less. Their intervention made a persuasive case that the reactor should be abandoned. Although the utilities did not make the decision to stop building the plant on their own, the New Hampshire Legislature later prohibited the pass-through of plant construction costs to ratepayers before the plant began operation. Unable to raise rates and strapped for cash to complete Seabrook 1, the utilities in 1982 were forced to abandon Seabrook 2.

This decision convinced CLF to begin intervening in the review of utility resource plans before PUCs throughout New England to force utilities to increase DSM expenditures as an alternative to building more power plants. In Massachusetts and Connecticut their pleas received sympathetic hearings from state regulators who encouraged local utilities to work together with CLF to investigate the potential for energy conservation.

CLF's successful interventions in individual power plant siting cases earned the respect of several utilities, including New England Electric Service (NEES). CLF executive director Doug Foy first met NEES CEO

Samuel Huntington at a 1986 forum on electricity. Huntington had given a presentation on NEES's experience with conservation that Foy thought underestimated the potential for reducing the demand for new power plants. Over lunch the two realized that a relation between the two organizations could be mutually beneficial in achieving their goals. It is this kind of direct one-on-one exchange between citizen activists and utility executives that would come to characterize utility reforms in New England.

CLF realized that for energy efficiency measures to be supported by investor-owned utilities they needed to be profitable. A conservation plan coauthored by NEES and CLF allowed the utility to share in customer savings from DSM. The plan was approved by state regulators and would later become a model for utilities in other parts of the country to increase their DSM investments.

After the New England Governors' Conference released *A Plan for Meeting New England's Electricity Needs*, in July 1987, CLF issued *Power to Spare: A Plan for Increasing New England's Competitiveness Through Energy Efficiency*. The latter report criticized the Governors' plan, noting that New England's electric rates were already 25 percent higher than the rest of the country. Another round of huge investments in large central power stations would only drive rates up further. CLF argued that uncertain demand growth, unpredictable fossil fuel prices, and volatile interest rates made a strategy of relying on new large plants unacceptably risky.

At that time, New England had a peak load of about 18,000 MW— roughly the output of 18 large coal or nuclear plants. If demand for power were to grow by 2.2 percent annually through the year 2000, peak demand would increase by 27 percent or the equivalent of five new coal or nuclear plants. CLF argued that rather than accepting these projections as reality, utility planners needed to look at how electricity was being used. They noted that most of New England's electricity was used for lighting, industrial motors, and space conditioning. Most of the expected increase in demand was being created by new construction, particularly new office and retail buildings. They found that most of the region's electricity was used by just a small proportion of the utilities' customers, mostly industries and commercial facilities like hospitals, schools, and offices.

Regional Power Planning

Power to Spare made the case that new technologies and superior building design could significantly reduce the need for new electricity supplies in New England. New lighting equipment alone could save 70 to 80 percent of existing lighting electricity consumption in many buildings. Likewise, high-efficiency industrial motors and computerized motor controls could lower industry consumption by almost 20 percent. Improved insulation could reduce annual residential heating requirements by at least 40 percent.

CLF and its public interest group allies put forward a comprehensive action plan for the implementation of cost-effective energy efficiency measures. Many of CLF's recommendations could be implemented by individual utilities. Yet a regional approach would be needed to remove larger energy efficiency market barriers. Unlike other parts of our nation, New England's electric generation system is already tightly integrated. All regional power sources are scheduled and dispatched as if the region were controlled by a single utility. Consequently, when utilities in one part of the region fail to implement cost-effective energy efficiency investments, customers in other areas pay for a higher cost regional power mix and a larger reserve margin.

Regional supply planning in New England had largely been done by the New England Power Pool (NEPOOL), a consortium of the public and private utilities that produce most of the power in the region. Historically, NEPOOL advocated construction of large power plants, including Seabrook, to assure reliable regional power supplies. As a private entity, however, there was no direct public oversight of NEPOOL's resource planning. Public review of utility investments and resource plans is carried out by each of the six states in the region. This fragmentation of responsibility, according to CLF, had often resulted in policies that hindered planning for electric services at the lowest possible cost for the entire region. CLF advocated creation of a regional power planning council to study the region's electric needs. This regional approach to IRP was to guide CLF's activities for years to come.

By 1992, just five years after the warnings of an energy crisis on the Eastern Seaboard, concerns shifted to what to do about a large power surplus. Energy efficiency measures had, in fact, reduced demand; new cogeneration plants were quickly built in response to the governors' call for new power sources; and Seabrook had come on line. In addition, New England's economy dipped in the early 1990s, which lowered the overall demand for electricity in the region.

CLF faced calls to dismantle the DSM programs that had been established just a few years earlier. CLF understood that the programs could no longer be justified as a way of avoiding the construction of new power plants. With a large power surplus, no one was planning to build additional power plants anyway. In 1992, CLF released an update of its *Power to Spare* report. It reviewed the DSM programs that had been put in place over the previous five years and suggested that there were still reasons to keep them in place.

CLF argued that energy efficiency programs had helped New England industries compete and expand, benefited low-income customers, and reduced environmental compliance costs. They pointed out that energy efficiency investments also had the potential to create new jobs in New England. The region could become a major exporter of new DSM technologies to other parts of the world, such as Eastern Europe, that are searching for cleaner and less expensive energy sources.

Yet another benefit of these DSM activities was cleaner air. Much of New England is in serious violation of federal clean air standards. In parts of Massachusetts, Connecticut, and Rhode Island, ozone—the most toxic of regulated air pollutants—exceeds federal standards by as much as 35 percent. Under the Clean Air Act, the southern New England region must come into compliance or face both the loss of federal funding for transportation and restrictions on economic growth. CLF argued that reduction in air pollution from utility energy efficiency programs could save New England up to $1.6 billion in Clean Air Act compliance costs compared to the cost of implementing power plant stack controls (see Figure 5-1).

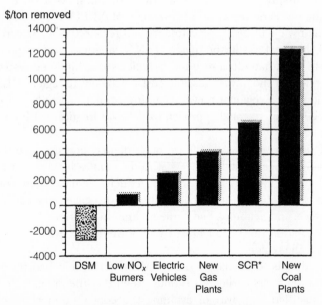

*Selective Catalytic Reduction on existing power plants.

Figure 5-1: Estimated costs of various NO_x control strategies. Utility energy efficiency programs are among the least expensive smog clean-up options available to New England because they provide substantial economic benefits to ratepayers apart from pollution control. New England utility DSM programs currently scheduled for implementation in this decade will save the region up to $1.6 billion in smog control costs alone. *Source:* Conservation Law Foundation, *Power to Spare II.*

East Coast Restructuring

The New England economy has since recovered. CLF was successful in preventing energy efficiency programs from being gutted during the recession, despite the fact that state energy policymakers were very concerned about the region's excess capacity. However, rates remain higher in New England than in other parts of the country. Some of the first calls for restructuring the electric services industry came from this region.

Large industrial customers argued that they should have access to the low-cost electricity available in the wholesale power market. As surpluses emerged in New England, short-term wholesale power transactions became more common. Companies like the big defense contractor Raytheon were among the first voices calling for "retail wheeling" in order to have direct access to the wholesale power market.

As in California, New England environmentalists initially strongly opposed retail wheeling. Armond Cohen, CLF's senior attorney and energy project director, wrote a trenchant analysis of retail wheeling for *The Electricity Journal* in April 1994. He persuasively argued that retail wheeling was a response to differences in rates resulting in large part from nuclear construction programs. Cohen noted, with some irony, that these ambitious construction programs, which created excess and underutilized power plant capacity, typically happened in areas that did not pursue IRP.

Cohen made the case that advocates of retail wheeling were still fighting "the last war—the war over nuclear imprudence. Sunk cost-shifting, not prospective cost-savings, is their end. Political muscle is their means." Cohen's article was published the same month that the CPUC issued its "Blue Book" retail wheeling order. The debate over retail wheeling took off in New England as well as in other parts of the country.

CLF came to recognize that the political muscle behind retail wheeling was considerable. More importantly, they saw the opportunity to use restructuring to accelerate the replacement of older, dirtier power plants with cleaner more efficient gas-fired power plants. Networking with other public interest groups throughout the country, CLF realized that the uncertainty that had descended on the utility industry after the release of the California "Blue Book" was constraining new investment in cleaner power sources.

CLF believed that new investments in low-cost combined cycle gas-fired power plants could displace nuclear and fossil fuel capacity. They decided that a retail wheeling world was better than uncertainty or political stalemate. Joe Chaisson, a CLF consultant, analyzed the air emissions from the current "fleet" of NEPOOL power plants and found that a substitution strategy utilizing new gas-fired plants would reduce sulfur dioxide emissions by 98 percent, NO_x by 94 percent, toxics by 99 percent, and carbon dioxide by 17 percent. CLF concluded that retail wheeling that resulted in replacement of existing power plants could actually be beneficial for New England's air quality.

Moreover, CLF's utility ally, NEES, saw that an early retail wheeling regime might give it a competitive advantage over other utilities in New England. The region's largest utility, Northeast Utilities of Hartford, Connecticut, had far more of its capital tied up in high-cost nuclear power plants than did NEES.

A key element to CLF's support for retail wheeling is a requirement that all old utility plants be brought up to the environmental standards that their new competitors must face. Some older fossil fuel plants emit 20 times as much air pollution as a new state-of-the-art power plant. This approach to creating a level playing field could clean up the air at a very low cost. The sticking point on this issue is a definition of how far and wide the "clean-up" rule for old sources of pollution would apply. If the area is defined too narrowly, then dirty plants in New York or Pennsylvania might not be covered. If the definition is too broad, then implementation becomes more difficult.

As in California, the most contested issue in New England is stranded cost recovery. The magnitude of stranded costs vary widely among the different New England states. The model developed by CLF and NEES is not based on the "lost revenues" approach endorsed by the CPUC and FERC but rather on the premise that utilities should have a reasonable opportunity to recover "net, non-mitigatable" costs arising from past decisions. While many policy analysts claim that this approach is preferable to a lost revenues approach, there is still considerable debate over the definition of "sunk" costs. For example, the Boston-based Competitive Power Coalition, which comprises independent power producers and power marketers, claimed that all of the investor-owned utilities serving Massachusetts, including NEES, include future costs for nuclear and fossil fuel facilities in stranded cost recovery proposals.

One of the more controversial approaches to stranded costs has been taken by the New Hampshire Public Utilities Commission (NHPUC) in conjunction with a two-year pilot project that allows a small percentage of the state's consumers to experiment with direct access purchases. Though the NHPUC staff has been dealing with stranded cost recovery proposals on a case-by-case basis, the overall policy is a 50-50 split between ratepayers and shareholders for stranded investments.

CLF estimates that the total stranded cost liability for the five states that comprise New England—Maine, Massachusetts, Rhode Island, New Hampshire and Vermont—is from $5 to $15 billion. These estimates vary widely due to the uncertainty about how these costs will ultimately be defined and assumptions about the price of electricity in future markets. This liability is spread unevenly among the states.

New England is working out the details of restructuring through a collaborative process involving utilities, large industries, and environmental organizations. The groundwork for this kind of cooperative relationship was laid during the struggles over Seabrook and CLF's interventions before state regulatory bodies over utility resource plans. State regulators in New England now look to the stakeholders to initiate cooperative solutions on energy issues. In some large stretches of the nation, however, state-regulated investor-owned utility monopolies are not the dominant providers of electricity.

The federal government created several regional power development corporations, two of which have had an immense influence on national energy policy and the people living in the regions they serve. The two areas in which the federal government played the leading role in developing electric power were the Pacific Northwest and the Tennessee Valley. Both regions embraced ambitious nuclear power programs influenced in part by the presence of large nuclear weapons installations and by the rapid growth each area experienced beginning in the 1960s. Both regions found atomic energy to be their nemesis.

The Pacific Northwest

The Pacific Northwest initiated its nuclear power program after the region's hydroelectric resources were largely developed. The successful taming of the region's rivers led to an attitude that "bigger was better." Building nuclear power plants would turn out to be very different from building dams.

The Pacific Northwest is defined by the colossal Columbia River basin. Its tributaries wind north through British Columbia and Alberta, east to the edges of Wyoming, and south to Nevada—a total of 1,214 miles. Harnessing the huge amounts of energy the Columbia River offered was part of a national mobilization leading up to World War II. During that time the Bonneville Power Association (BPA) even hired itinerant folk singer Woody Guthrie to capture in music what the federal agency was all about. One of his tunes was called *Talking Columbia:*

> *"Folks need houses and stuff to eat,*
> *And folks need the metals and the folks need wheat,*
> *Folks need water and power dams,*
> *Folks need people and the people need the land,*
> *The whole big Pacific Northwest up here ought to be run,*
> *way I see it, by electricity. . . ."*

BPA ultimately built 30 power stations on the Columbia. The large power houses turned bauxite into aluminum at huge smelters along the river. This aluminum was taken to factories in Washington and shaped into fuselages and wings for planes for the U.S. Air Force. The aluminum smelters were typically located right next to the power stations completed at Bonneville Dam in 1938, the Grand Coulee in 1941, and other sites. More than a dozen of these huge industrial outfits became direct customers of BPA, an arrangement that would later affect the debate over restructuring.

In 1943 the aluminum industry consumed 60 percent of the power Bonneville sold. By the 1950s, most of the Columbia River system within U.S. borders had been developed for power. In 1964 Canada and the United States signed the Columbia River Treaty, which added new hydroelectric capacity in British Columbia and developed transmission links between BPA and BC Hydro, the Canadian provincial government utility. As the 1960s drew to a close, however, BPA and others began to look beyond the river for new energy sources.

Washington State is among a handful of states that embraced the notion of local control of electricity supply. The state's public power laws encouraged the creation of a host of rural electric cooperatives and municipal utilities, most managed by locally elected boards of directors. The state's two largest cities, Seattle and Tacoma, are served by municipal systems. However, most of the public power systems in the state were quite small.

The Washington Public Power Supply System

In 1957 seventeen small, mostly rural public utility districts banded together to create the Washington Public Power Supply System (WPPSS). This consortium gained the status of a "joint action agency," which under state law was allowed to develop new electric generation facilities.

The region's first foray into nuclear power was a special reactor to generate electricity at the Hanford Reservation in eastern Washington near Richland, a sprawling complex that supplied plutonium for the nation's nuclear arsenal. An 860-MW plutonium reactor was completed at the site in 1966. WPPSS originally intended to upgrade this plutonium-production reactor for electricity production. However, it was persuaded instead to build a new 1,100-MW General Electric reactor also to be located at Hanford. When this reactor was ordered in 1974, its estimated cost was $398 million. It is the only reactor that WPPSS completed, and the final cost was over $3 billion. Later on, WPPSS tried to build four more reactors. None were completed, but they were not canceled until they consumed $13 billion.

By 1980 WPPSS had grown to include 70 publicly owned utilities. The organization also defaulted on $2.25 billion worth of bonds it had issued for the construction of two nuclear plants. This default, the largest ever for a municipal government agency, was part of a larger regional plan that at one time called for the construction of 26 large coal and nuclear power plants. It was one of the most dramatic examples of the failure of traditional utility planning methods.

WPPSS's governing board evolved into a 70-member committee. According to critics such as James Leigland and Robert Lamb, authors of *Who is to Blame for the WPPSS Disaster*, it was the inexperience of many of these decisionmakers, as well as the unmanageable size of the governing body, that allowed these reactors to go forward despite increasing evidence that choosing nuclear might be a mistake. However, much of the blame for the fiasco lies with the nuclear contractors who kept bad news from the governing board, the financial community that continued to sell bonds that put the agency further in debt, and the utility managers who were unwilling to look at alternatives to the nuclear program. But the fundamental problem was lack of public accountability. While the citizens at some local public utility districts tried to get out of WPPSS, the complex nature of the joint powers agency made that impossible. There was no mechanism by which the public could put forward alternatives to the nuclear project. In 1980 WPPSS defaulted on the debt for two nuclear plants. Still, they continued with the program to build the other three nuclear plants.

In 1981 Washington voters approved a statewide referendum which stated that future WPPSS bond sales would have to be approved by a public vote. That same year, BPA decided to mothball two of the remaining reactors. However, until 1994 the federal agency continued to fund skeleton operations at the two partially built plants to keep them ready for completion sometime in the future. Finally, in 1994, after a plan to use the reactors to burn plutonium fizzled, Bonneville decided to scrap the reactors. Whereas the rates charged by BPA steadily declined between 1940 and the late 1970s, the costs associated with WPPSS projects later forced large increases in the rate BPA charged its residential customers (see Figure 5-2).

The WPPSS fiasco led to the passage in 1980 of the Pacific Northwest Electric Power Planning and Conservation Act, which empowered BPA to acquire new energy resources. According to the U.S. Congressional Office of Technology Assessment, the Act represented "a unique attempt by the Congress to encourage the Pacific Northwest region to set a national standard in determining the wise use of limited resources, protecting the environment, ensuring equitable distribution of the costs and benefits of power needs, and testing the opportunities for shifting conservation and renewable resources to provide a stable and substantial future."

The nation's first regional power supply planning body, the Northwest Power Planning Council (NPPC), was created by the Act. This new council was instructed to conduct regional reviews of power demand and supply for the four states in the Columbia River basin: Washington, Oregon, Mon-

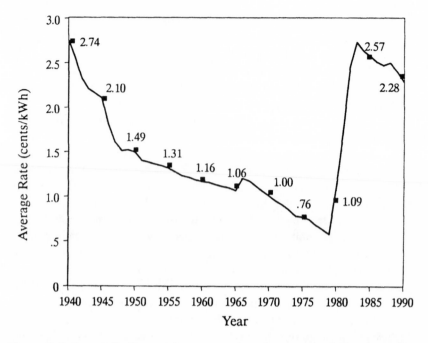

Figure 5-2: Bonneville Power Administration's preference rate declined steadily until 1980, when WPPSS nuclear investments reversed a steady decline in BPA's preference rate for residential consumers. (All figures are in 1990 dollars adjusted for inflation.) *Source:* Northwest Power Planning Council, *1991 Northwest Power Plan*, Volume II.

tana, and Idaho. Among the duties of the NPPC were preparation of a 20-year regional power plan and development of a fish and wildlife program to protect and enhance native salmon and fish populations that had been devastated by the region's hydroelectric power houses. The NPPC was also mandated to seek broad public participation in these processes.

The Act established a set of priorities for new resource acquisition. Conservation came first; non-hydro renewables (wind, solar, biomass, geothermal) came second; low-pollution sources, including gas-fired cogeneration and fuel cells came third; and traditional sources such as coal and nuclear now sat at the bottom.

The door for citizen involvement in IRP was also opened by the Act. Local citizens seized the initiative. In 1982 the Northwest Conservation Act Coalition (NCAC), an umbrella group for more than 65 public interest organizations and progressive utilities, developed a sweeping "Model Electric Power and Conservation Plan." This model plan set a framework for the

NPPC's first plan released in 1983. NCAC also stimulated the development of a Model Action Plan (MAP) and led a public education campaign that delivered more than 1,500 written and oral comments to the NPPC. The vast majority of these comments endorsed conservation and renewable resources and opposed further exploitation of coal or nuclear fuels. Themes developed in this MAP shaped the NPPC's 1991 regional supply plan.

BPA also launched one of the nation's first major conservation programs, investing $21 million to retrofit 90 percent of the residences of Hood River, a town founded by fruit growers in the 1850s. This pioneering effort, launched in the spring of 1982, was based on a thorough assessment of one community to identify barriers to energy efficiency among different social groups. This research set the stage for a successful campaign to implement energy efficiency measures that provided valuable information for subsequent conservation efforts in other parts of the country.

Among the other major accomplishments of NCAC and regional environmental groups, working in cooperation with forward-looking utilities, was regulatory reform that decoupled utility revenues from increased sales for investor-owned utilities such as Puget Power & Light. Also adopted were strong residential and commercial energy efficiency codes that supplemented utility DSM programs and demonstrated the role that conservation could play in a utility-dominated market.

Trojan

The debate over restructuring the Pacific Northwest's energy services industry was also shaped by citizen efforts to close another nuclear plant, the 1,100-MW Trojan reactor operated by Portland General Electric (PGE). The plant was constructed on the banks of the Columbia River near Rainier, Oregon, in 1975.

As Trojan aged it was plagued by persistent leaks in its steam generators. PGE spent millions of dollars to plug holes in hundreds of the heat transfer tubes that were used to flash steam to power a steam turbine. The microscopic fissures enabled radioactive contaminated water to seep into the secondary heat loop that provided the steam to generate electricity. Once the level of radioactivity in this loop got too high, Trojan had to be shut down and cleaned up.

During the 1980s, the citizens of Oregon had tried twice to close Trojan through statewide ballot initiatives. PGE had defeated both with well-financed campaigns. However, after the successful effort by Sacramento voters to close Rancho Seco, the region's safe energy activists decided to try again. Two measures were placed on the November 1992 ballot. One was sponsored by the grass roots activists responsible for the two previous campaigns. It simply called for Trojan's closure. The other was sponsored by Jerry and Marilyn Wilson, owners of Soloflex—a manufacturer of body building equipment. The Wilsons contributed over $1.5 million to their initiative calling not only for Trojan's closure but for shifting decommissioning costs to PGE shareholders.

This latter provision troubled PGE and was seen as unfair and possibly illegal by parties that were otherwise sympathetic to Trojan's closure. The cost of decommissioning a nuclear plant is incurred at the time the reactor is contaminated, that is, when it is started up. Organizations like NRDC, which agreed that Trojan should be phased out, were concerned that passing an initiative that prevented PGE from collecting decommissioning funds from ratepayers sent the wrong signal to utilities that were considering the early retirement of nuclear plants. They argued that utilities should not be penalized for making an economically sensible decision.

Early on during the anti-Trojan campaign, PGE convened a citizen's advisory group to look at alternatives to the nuclear plant. The group was broad-based and included several environmental organizations. In collaboration with PGE management, the group came up with a plan that would phase the plant out in four years, at the time when the huge steam generators would need to be replaced. For groups like NRDC and the NCAC, this seemed like a reasonable compromise.

However, many safe energy activists did not trust PGE. They saw the phase-out proposal as part of the campaign to defeat the ballot initiatives. Also, they were concerned with the manner in which a nuclear plant would be operated under the equivalent of a death sentence. Since the plant would be closed in the near future, they argued that workers would be concerned about looking for new jobs and management would be less willing to invest in plant safety measures.

PGE spent $5 million to defeat the two ballot measures. However, one week after the election, Trojan was taken off-line because of more leaks in the tubes in the steam generators. Immediately following the shutdown, the Union of Concerned Scientists discovered NRC documents confirming that microscopic cracks in the tubes were more serious than PGE had led the public to believe.

The utility was in a quandary. It just won the election to keep the plant open for four more years, yet the credibility of its statements about the reactor's safety were now seriously challenged. The utility initially wanted to repair the leaking tubes and return the plant to operation. However, the NRC would require a large-scale inspection of all the tubes in the four steam generators.

Like the situation at SMUD, there were factions inside PGE that thought it would be prudent to immediately shut down Trojan. After the latest revelations, they prevailed. PGE's chairman, Ken Harrison, announced in January 1993 that closing Trojan was the "least cost decision" for the utility. He argued that new information about the availability of low-cost energy in the wholesale power market had led to the decision. Later, the utility calculated that the early closure of the plant would save Portland's ratepayers more than $800 million over the next 20 years. PGE announced its closure of the reactor before it had received any public assurances from the Oregon PUC that its investment in Trojan would be recovered. The PUC decided to allow PGE to recover from its ratepayers 86 percent of its outstanding investment in the plant, as well as all the decommissioning costs. PGE shareholders would have to cover the remaining 14 percent of Trojan's debt.

K.C. Golden, former head of the NCAC, was one of the environmentalists that sat on the panel that looked at PGE's options. "The pressure of the initiatives was very constructive," said Golden, noting that they got PGE to the point where the utility was willing to seek a face-saving way out of its nuclear dilemma. Rather than go "for total surrender and humiliation," Golden reasoned that using economic arguments for closure would set a better precedent. "One of the opportunities associated with increased competition is to retire uneconomic and environmentally unacceptable units," stated Golden. He thought Trojan could serve as a model for other utilities looking for a way out from under the burden of nuclear liabilities.

The following January, a coalition of environmentalists and public interest organizations presented PGE with a model resource replacement strategy. The utility wanted to boost its use of natural gas for generation of electricity from 17 percent to over 50 percent by the year 2001. Environmentalists suggested instead that the utility should increase its investment in renewable energy sources. After the Trojan fight, PGE had learned the value of listening to its customers. They agreed to release a Request for Proposals earmarked for renewable resources to help replace the output of Trojan. However, the utility, like its counterparts in the rest of the country, instituted deep cuts in conservation spending in 1996.

Renewables Northwest Project

With the creation of the Northwest Power Planning Council and the opening up of PGE to more citizen participation in decision making, the citizens of the Pacific Northwest had generated opportunities to remake the region's energy future. And, indeed, the NPPC had developed a long-term plan that had been heavily influenced by citizen involvement. However, it was one thing to have a plan and another to transform it into reality.

To address this gap between plans and implementation, the Renewables Northwest Project (RNP) was founded in 1993 to build support among local environmentalists that had become suspicious of all electric generating technologies, including renewables. Other groups targeted by RNP's public education campaign include businesses and opinion leaders. RNP advocates a sustained orderly development policy to stimulate non-hydro renewables, such as wind and geothermal, in the region. At present, six wind and geothermal projects representing 250 MW of renewable capacity are under development with the help of RNP.

RNP's strategy for stimulating renewables was modeled on the way the region had promoted recycling. "Recycling mandates established a base market with government institutions," noted Peter West, an RNP advisor. But after private sector institutions were established, market forces began to take over. The policy envisioned by RNP for renewables, which it refers to as "policy push/market pull," is one in which government sets a mandated minimum for renewables and then lets customers make additional investments in the percentage of "green kilowatts" they want in their resource mix through green pricing.

Low energy prices in the Pacific Northwest are both a curse and a blessing for RNP.[1] The bad news is that it is very difficult to compete against such low prices. The good news is that some consumers are willing to pay a little extra for clean energy because the end result is still a very competitive electric rate. The community of Salem, Oregon, for example, decided it wanted to become the nation's first community to be served solely by renewable power sources after a poll of the local community indicated widespread support for renewable technologies and energy conservation.

RNP executive director Rachel Shimshak emphasizes that the customers of Salem Electric, a municipal utility, strongly supported the notion that costs associated with the initial stages of renewable development in the Pacific Northwest should be spread among all consumers, as was the case with SMUD for its wind and biomass plants. She sees the model used by SMUD as being appropriate for the Pacific Northwest. "SMUD got it right," said Shimshak, expressing confidence that once a base market was established, consumers in the Pacific Northwest could select additional cleaner sources of energy if given the opportunity through a green pricing program like SMUD's PV pioneers.

Shimshak cites Salem Electric as a model for the future. The utility requested that none of the power it purchases from BPA come from the agency's nuclear plant. The utility is developing plans to use non-hydro renewables, such as wind power, to make up the difference. "Salem Electric may become the first utility in the country to purchase 100 percent of its energy needs from renewables," said Shimshak.

Pacific Northwest Restructuring

Unlike California, which initiated a series of public hearings after the state public utility commission issued its controversial "Blue Book" retail wheeling order, this region is holding hearings before a specific deregulation proposal is put on the table. A representative group of stakeholders from the four Pacific Northwest states was invited to begin the deregulation debate. The participants organized themselves as a learning group. At biweekly sessions they conduct exercises that help them envision new market structures and planning procedures.

The push for competition in the Pacific Northwest is coming from those industrial customers who do not have direct access to the wholesale market. The aluminum smelters that do, referred to as "direct services industries (DSI)," have recently discovered that they could, for the first time, access cheaper power than what BPA was offering. The DSI customers represent one-third of BPA's load. Power purchase contracts recently approved by BPA allow these large DSI customers to escape the agency's stranded nuclear costs while still allowing them to use BPA transmission lines to import power.[2] That decision led to protests from other customer groups and from the environmental community. A comprehensive review of the region's energy policy grew out of this controversy.

Chuck Collins, one of the original members of the NPPC, is heading up the comprehensive review. He noted that one area of consensus is around the need for an independent grid operator to "help consolidate twelve control areas into one transmission system." In the spring of 1996, the suggestion to separate BPA's transmission function from its generation responsibilities was also widely supported.

Although many public power agencies are resisting the notion that deregulation is necessary in the Pacific Northwest, Brett Wilcox of Northwest Aluminum Company has put forward a reform plan for BPA that he calls "de-federalization." He proposes to transfer the agency's physical assets to state governments. He cited the transfer of the Alaska Railroad and the largest hydro project of the Alaska Power Administration, both federal facilities, to the Alaskan state government as a model that could be applied to BPA.[3]

Under Wilcox's scenario, BPA's power marketing role would be turned over to a new nonprofit entity owned by BPA's former customers and other regional purchasers of bulk power. Specified portions of sales revenue could still be earmarked for social programs like fish restoration on the Columbia River. The proposal also calls for an independent grid operator that would provide nondiscriminatory open access transmission service. Though environmentalists find some of Wilcox's suggestions thought-provoking, they are adamantly opposed to any deal allowing industrial consumers to escape responsibility for the region's nuclear debt. Another approach being examined by the NPPC is to break out the WPPSS debt from electric rates and allocate these costs fairly among customers large and small.

The amount of the region's nuclear debt can be seen by examining BPA's 1994 fiscal year budget. The nuclear program represented $737.1 million or 32.7 percent of the agency's total costs that year. Some $386 million are expenses associated with WNP2, the lone operating reactor; the remaining $350 million in costs is linked to three reactors that have been scrapped: WNP1, WNP3, and Trojan.[4]

Among the options suggested by the NCAC for dealing with these stranded costs is the imposition of an exit fee on large consumers who wish to abrogate current power purchase contracts and have direct access to the wholesale market. Another approach would be for BPA not to renew power purchase contracts with DSI customers in 2001 if they do not accept their share of the region's stranded costs. The DSI load will then be served by higher priced electricity marketed by investor-owned utilities. In this circumstance, losing the DSI load saves BPA money, because it would not have to purchase higher cost power to meet the high nighttime loads of these industries. The low-cost hydro power would be preserved for meeting intermediate and peak loads.

The role of BPA in a more competitive electricity market may become even more important in managing the region's power supply. To date, BPA's reaction to the forces of competition, however, echo that of investor-owned utilities. Its DSM budget was slashed by 90 percent and funding to assist in the restoration of native salmon runs has also been cut, despite widespread agreement that such programs are popular. Support for several renewable pilot projects developed in conjunction with the RNP are also jeopardized.

Some argue that BPA should be stripped of its social purposes and act more like an ordinary utility. Many local activists disagree, claiming that BPA can accomplish all of its mission, including restoration of native fish populations, if the stranded cost issue is addressed equitably.

The Columbia River can produce abundant salmon and steelhead, states K.C. Golden, and it can provide irrigation and power. It can also help support new ventures in energy efficiency and renewable energy development. "The Columbia River can do a great many marvelous things, but the one thing it *cannot* do is to continue to act as the limitless financial sponge for the unproductive WPPSS debt that BPA incurred on its customers' behalf," said Golden in testimony filed in the spring of 1995. "If BPA is to survive as a uniquely valuable public institution, it cannot become the vehicle into which BPA customers pack their nuclear obligations and ship them, COD, to the Federal Treasury," he added. Golden suggested fees collected outside of BPA's wholesale power rate, either on the transmission or distribution system, as a rational way to address the stranded cost issue.

The Tennessee Valley

The Pacific Northwest has a long tradition of citizen activism. The other part of the country with a large federal electric system does not. Only recently have members of the public begun to question the authority and the vision of the Tennessee Valley Authority (TVA). Alex Radin, longtime executive director of the American Public Power Association, is a keen observer of public power agencies across the country. He described the difference between the TVA and BPA as follows: "The culture of the agency is important but the culture of the region is also important. The culture of the Northwest is so different from Tennessee Valley culture. There is a lot more public input in Bonneville in that the people of the area are more concerned about the environment and what the agency does. If these environmental issues were of greater concern to the people of the Tennessee Valley, TVA would be more responsive. TVA's open board meetings are not real scrutiny. Bonneville has a series of meetings throughout the region in which they lay out their plans. There is a lot more activism there and also more give and take."

TVA launched America's most ambitious nuclear power construction campaign, which grew to include 17 reactors under construction or in operation until it collapsed in 1985 under a growing mountain of debt. TVA has the distinction of bringing the nation's last reactor on line.

TVA seemed to be an ideal organization to commercialize nuclear power. It was founded in 1933 in the midst of the Great Depression in an area left behind by the wave of industrialization that had rolled across the country. Its founding leader, David Lilienthal, was a visionary who imbued the organization with a large sense of mission: electrify the Tennessee Valley, an area consisting of the state of Tennessee and parts of six other states. By 1955, just 20 years after embarking on this economic development effort, every farm and home in the valley was wired for electricity.

The TVA took enormous pride in its accomplishments. It had a "can do" spirit that had enabled it to build and operate some of the largest power plants in the world. There was no reason why it couldn't build nuclear power plants as well or better than any other utility. Besides, the Oak Ridge National Laboratory, which had played an instrumental role in the Manhattan Project, was located in Tennessee. It could provide expertise and training in nuclear matters to the TVA staff. The people of the valley were also used to experiments with nuclear fission and saw it as an economic boon.

Until 1979 no one dared question the wisdom of TVA's massive investment in nuclear power. The Congressional General Accounting Office (GAO) that year strongly criticized the agency's forecasting methods and argued that several of the nuclear plants under construction were not needed. The TVA board deferred construction of four reactors. Still, the nuclear program escalated out of control. The proposed 17 reactor system was originally estimated to cost $7 billion; by 1980 the price tag had inflated to $17 billion.

Beginning in 1980, these nuclear investments forced TVA to impose drastic rate increases. Within 18 months, the TVA board voted to raise rates three times, leading to rate increases totaling 36 percent. And by 1984, the TVA board canceled four more reactors, leaving five licensed plants and four more under construction. By that time, the operating reactors began to have problems. That same year, the two Sequoyah reactors were closed because of safety problems. A year later, the three Browns Ferry reactors were also shut down indefinitely. Twenty years after the TVA went nuclear and invested $14 billion, not one reactor was running.

The TVA is unusual among American political institutions in its low level of accountability to elected officials. Though its three-member board is appointed by the President of the United States, its power program is self-financing. Unlike BPA and other federal power marketing agencies, it is not directly accountable to Congress for funding. It can set its own rates, so there is no public oversight by way of a PUC. Wall Street assumes that its debt is backed by the full faith and credit of the U.S. government, so it is not overly concerned about bad investment decisions. The only realistic check on the TVA's power is its debt ceiling which must be authorized by Congress. However, in 1979 Congress doubled the debt limit from $15 to $30 billion with hardly the blink of an eye.

In 1988 President Bush appointed Marvin Runyon, the past president of Nissan Motor Manufacturing Corporation, USA, as chairman of TVA. Runyon made major cuts in the TVA work force, laying off 7,500 people and eliminating another 5,000 jobs through attrition. Overall, employees were reduced by more than one-third. He slashed most of the forward-looking conservation programs that S. David Freeman had started a decade earlier when he was TVA board chairman. However, four uncompleted reactors were still consuming large quantities of capital.

In 1992 President Clinton appointed Craven Crowell, a journalist who had worked at the TVA as director of public information, as board chairman.

One of Crowell's first acts was to invite the public to participate in the creation of a long-term integrated resource plan. Genuine public participation would be a new experience for the TVA, which, despite being a government agency, had operated for decades like a private club. It would also be a challenge to the political culture of the Tennessee Valley, where strong public oversight of electric utilities had not been a tradition.

In 1995 the TVA staff prepared a document entitled *Energy Vision 2020* outlining a 25-year long-term energy plan and a seven-year short-term action plan. The short-term action plan recognized that significant changes were occurring in the electric services industry. It recommended completing only one of the partly built nuclear reactors—Watts Bar. Though an admission of past nuclear mistakes was welcome, the IRP was not able to make a clean break from the past. Watts Bar was a source of major controversy. The NRC discovered major welding problems that delayed construction for years and raised questions about the competence of TVA management. When the IRP plan was released, Watts Bar had been under construction for 23 years and had run up a bill of $6.2 billion. With the cancellation of the other TVA reactors, Watts Bar was the only nuclear plant still under construction in the United States. The IRP plan also recommended that the TVA study repowering the abandoned Bellafonte nuclear power plants using natural gas or gasified coal, a strategy that had already proved very costly in Michigan and Ohio and that had been rejected by SMUD.

The underlying issue that was weaved in and out of the IRP plan was the need for the TVA to respond to the market and to political pressures brought on by an increasingly competitive wholesale power market. The larger cities in the Valley, like Memphis, Nashville, and Knoxville, knew the bargains that other municipal utilities, like SMUD, were getting in the wholesale power market. Large industrial companies like Tenneco were also aware of trends in other parts of the country. TVA's rates were still competitive with neighboring utilities. Yet once it started paying off its nuclear debt, it would have to raise rates. The sooner it began paying off the debt, the more flexibility the agency would have in the longer run.

Still, there were legal and structural incentives for TVA management to ignore the nuclear debt problem. TVA's policy was not to have ratepayers pay for construction work in progress. This policy, in itself, was reasonable. In fact, most states do not allow utilities to collect funds invested in a power plant from ratepayers until the plant is used. But for years it was apparent that most, if not all, of TVA's partly built nuclear power plants would never be finished. If the TVA admitted this, it would constitute plant abandonment. In that case, a utility has two choices: one is to write down shareholder equity; the other is to get the ratepayers to pay off the plant's debt. For a public utility like SMUD—with no shareholders—the ratepayers bear the full burden of paying off the nuclear mortgage. The TVA could choose to have their ratepayers pay for the abandoned plants or convince Congress to let federal taxpayers finance the debt.

TVA management did not like either of the options, so it maintained the fiction that the nuclear power plants would be finished one day. No one questioned this strategy until 1995, when the GAO issued another scathing report saying deregulation of wholesale power markets would ultimately cause TVA to lose its monopoly status. Then it would not be able to raise rates to pay off the debt on the abandoned nuclear plants. U.S. taxpayers would then be stuck with the bill.

The TVA enjoys a unique monopoly status among wholesale power providers. Federal law prevents its customers from buying power from anyone else. This law gives TVA management the confidence that it will not lose customers to other wholesale power suppliers. However, if the agency raised rates and refused to allow the cities and other local municipal utilities to buy lower cost electricity elsewhere, the TVA would be in a difficult political situation. Local environmentalists thought the TVA might be able to use its monopoly status to reduce long-term risk, assuring that it would be competitive in the future. If the TVA could take advantage of the low prices offered in the contemporary regional wholesale power market, it might be able to begin paying off its nuclear debt. That is the strategy that SMUD has used. To follow this path, TVA would have to become a net importer of power. That went against everything that the TVA had stood for in the past. The utility had always intended to export, not import, power.

Local environmental groups, led by the Tennessee Valley Energy Reform Coalition (TVERC), were concerned about the safety of Watts Bar. Whistleblowers at the plant suggested that major problems remained unsolved. If the reactor started generating electricity, they thought it was unlikely to run well and would soon require costly repairs. These environmentalists asked TVA to cut its losses and reduce future borrowing by terminating the start-up of Watts Bar and look instead at how SMUD acquired new resources, suggesting that TVA identify cogeneration project opportunities.

TVA should use DSM programs as part of a strategy to build customer loyalty and to establish long-term agreements, making the TVA the preferred provider of electricity for the valley's customers, the environmental groups argued. Citizen groups also wanted the TVA to invest more in PV, wind, and biomass technologies. The agency's IRP process was a bona fide effort by the agency to reach out to various constituencies—both regionally and nationally—concerned with energy policy. To its credit, TVA formed a stakeholder review group involving a diverse array of citizens to design a new resource strategy. After a draft IRP was prepared, nine public meetings were held. These hearings, nonetheless, were adversarial because TVA was unwilling to consider alternatives to starting Watts Bar.

Watts Bar was finally fired up in early 1996, the last nuclear facility to go into service in the United States. How well it runs will greatly affect the TVA's future. Problems in the early years requiring expensive upgrades will place TVA in a precarious financial situation. After the reactor start-up, the TVA board adopted *Vision 2020*, which expanded the agency's commitment to DSM by 750 MW over the original draft proposal. By 2010, the agency plans to add 2,200 MW of new DSM measures. In addition, the TVA agreed to investigate the potential for a wind power project in its service area and to study the possibility of building a biomass facility. The upgrades were applauded by environmentalists. Still, many questioned the vagueness of the commitment to renewable energy and the decision to continue studying the conversion of the unfinished Bellafonte nuclear plant to use natural gas or gasified coal.

Steve Smith, TVERC director, complains that though TVA is at least acknowledging the value of clean power sources such as wind power, the agency is still "unwilling to take on any new risks." He adds: "The black hole of this region's nuclear legacy was so destructive. It is sucking everything good at TVA into it." Smith predicted that TVA will reach a crisis within the next few years due to a nuclear stranded cost figure that stands at $27 billion, and counting. "The real issue here is the hot potato of nuclear stranded costs," said Smith, noting that as soon as any of TVA's 160 wholesale customers bolt to go their own way in a deregulated market, rate increases will follow. His prescription for reform? TVA should sell off its nuclear and coal assets and return to its original mission—the management of rivers and dams. He also suggested that any new capacity needs identified by TVA be earmarked for renewable resources.

Lone Star Independence

The state of Texas started out as an independent republic. At times, it still acts like a nation unto itself. This attitude is reflected in the evolution of the state's energy policy.

Like the Pacific Northwest, large parts of rural Texas first received electricity through the rural co-op public power movement. A young Congressman, Lyndon Baines Johnson, a strong advocate of public power, helped secure funding to set up the Lower Colorado River Authority in the hills outside of Austin. The new power generation stations in central Texas greatly improved the lives of people living throughout the region and helped Johnson on his journey to the White House.

Of course, the most important reason for the early growth of the Texas economy was the discovery of oil. The wildcatter days at the turn of the century exemplify Texas's entrepreneurial spirit and attitude toward energy sources. They would also set the stage for the development of a unique electricity system.

For many Texans, the thought of running out of power was akin to losing a war. It was certainly better to have too much energy than too little. So Texans built and built. If all of the state's power plants were up and running, they would produce at least 35 percent more power than Texans need. Texas has been highly resistant to regulation of energy matters. It was the last state in the country to form a Public Utilities Commission, in 1975. In fact, two nuclear reactors constructed in Texas—the South Texas and Comanche Peak power plants—were never officially permitted, or put through any environmental review, by the Public Utilities Commission of Texas (PUCT).

Texas also has unique views on regulation by the federal government. It wants to avoid it at all costs. That's why most of the state's utilities formed the Electricity Reliability Council of Texas (ERCOT), a group whose members have pledged not to sell excess capacity beyond the state's borders in order to avoid regulation by FERC. By not engaging in interstate transactions, these utilities were not engaged in interstate commerce, the justification for FERC's jurisdiction.[5]

The passage of a 1978 federal law restricting the use of natural gas for electricity generation resulted in a building boom for coal and nuclear capacity in the state. Unlike California, which focused on renewable energy sources, Texas turned to these more conventional sources of power. In 1975 Texas was dependent on natural gas for 90 percent of its electricity. By 1987, the state's fuel diversity program brought this percentage down to 45 percent, primarily through the use of coal and lignite, which each account for about 25 percent of current Texas electricity generation. The remaining 5 percent is supplied by nuclear power. Texas's energy consumption reflects the more-is-better attitude with per capita energy use far greater than that in California and other states that have promoted energy conservation (see Figures 5-3 and 5-4).[6]

Though rates paid by Texans are not high—residential rates rank 22nd and industrial rates 37th in the country—bills paid by consumers are among the nation's highest. Policymakers have paid lip service to the value of energy efficiency as far back as 1983, but Texas utilities still use energy sales to boost revenue. DSM programs have not been a high priority among Texas utilities, with the exception of two public power entities—the Austin Municipal Utility District and the Lower Colorado River Authority. Karl Rabago, a former PUCT commissioner and an analyst with the Environmental Defense Fund (EDF), observes that Texas utilities have been averse

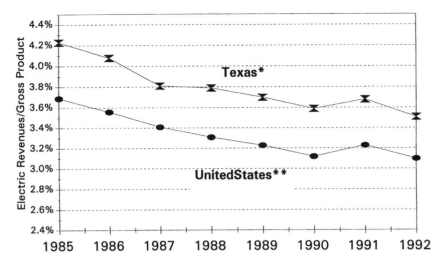

Figure 5-3: Electricity consumption in Texas is far above the national average. *Source:* Texas Ratepayers' Organization to Save Energy, Inc.

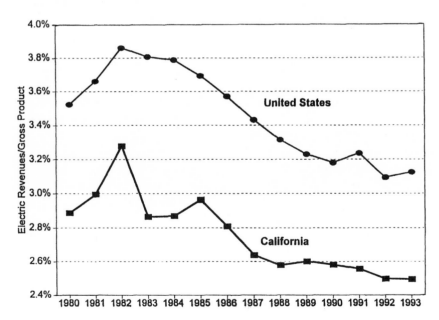

Figure 5-4: Successful conservation programs in California have reduced demand for electricity, making the state's economy far more efficient than states such as Texas. *Source:* California Energy Commission, *Quarterly Fuel and Energy Report* (for California data); National Energy Information Center, U.S. DOE (for U.S. data).

to the type of collaboratives used by environmentalists in other parts of the country because they had never been challenged on past decisions about nuclear and coal plants. "We never had them over a barrel," acknowledges Rabago.

One of the peculiarities of Texas's approach to energy policy is associated with cogeneration. Though the state ranks at the top when it comes to installed cogeneration capacity, only about half of the power generated at these facilities is sold to utilities. Although cogeneration plants are often less costly than utility power plants, state policy makes it difficult for the owners of these facilities to find a market for the electricity they produce. This policy results in higher rates for the state's residential ratepayers.

According to estimates made in 1989, Texas could economically develop 17,000 MW of cogeneration. As of 1991, some 8,295 MW was developed. By 1996, only 1,900 MW will have firm power sale contracts. "That means there will be 6,000 MW of cogeneration looking for a buyer," observed Tom "Smitty" Smith, an analyst with Public Citizen, a statewide environmental group. Trying to find alternative buyers of this power is now driving the effort to open up the Texas market, he said. "Refineries want to sell power at the retail level to surrounding communities. They want to make cogeneration systems their new profit centers," Smith added.

Clean Power Potential

Along with this potential for gas-fired cogeneration, Texas has abundant sun and wind resources. Texas reportedly has more renewable energy resources than any other state, except California. The wind blowing throughout the state could supply 10 percent of the nation's total demand for electricity, or over five times what Texans need. At present, however, Texas ranks at the very bottom of states in terms of renewable capacity currently on-line, according to report by EDF entitled *The Next Texas Energy Boom*. "Less than half of one percent of the state's energy consumption—one fifteenth the national average—is being supplied by renewables; only Kansas ranks lower," claims this report. However, the decline of the state's oil industry, as well as an aversion to importing power from outside of its borders, could now fuel a new renewable energy industry.

Already a number of utilities are investigating wind. New World Power has announced a 40-MW project with Texas Utilities of Dallas. Central and South West Corporation, also of Dallas, is building a 6-MW facility featuring a new turbine manufactured by Zond Systems. And Kenetech Windpower has plans to develop a wind project for the Lower Colorado River Authority.

The critical roadblock to innovations in Texas are past investments in nuclear and coal capacity. Three out of five utilities in Texas own a chunk of a nuclear reactor. Those served by the utilities owning nuclear capacity are paying two cents more per kWh than those served by nonnuclear utilities. And these costs are increasing. Problems continue to plague the 2,500-MW South Texas nuclear power station. A report prepared by Public Citizen shows that the South Texas reactor's O&M costs are 7.9 cents/kWh; while cogenerated electricity is available at about 2 cents/kWh.

The other key barriers keeping Texas from moving to a more sustainable energy path are the state's older coal plants. Like California, Texas is awash in excess capacity. Peak demand in Texas is approximately 52,600 MW. On-line generating capacity is 68,000 MW. Unless regulators allow new and cleaner sources to compete with coal plants using comparable environmental standards, large-scale deployment of renewable resources will be thwarted. The 11 oldest and dirtiest coal plants in the system represent 16,000 MW of capacity. Utility power plants, and oil refineries, are major sources of air pollution. More than 153 million tons of carbon dioxide spew from Texas utility plants and oil refiners. Though refinery contributions are larger at 44 percent, the utility power plant share of 29 percent is a major contributor to pollution in major population centers, such as Dallas and San Antonio.

Ironically, because Texas was so tardy in adopting state electricity regulations that are common elsewhere, its utilities are only now moving forward with their first full-fledged integrated resource plans while simultaneously contemplating a major restructuring of the electricity marketplace. Though Texas has flirted with IRP as far back as 1983, its efforts have been superficial at best. Nonetheless, after three failed attempts, the Texas Legislature finally approved IRP legislation in 1995. Because this IRP law is so recent, it contains rules on competitive solicitations that are specifically designed to meet the needs of the emerging wholesale market. The main value of the IRP, however, will be to create a public process for considering which if any of the over 60 generating facilities now on the drawing boards should be developed.

Though the historic *laissez-faire* approach to utility regulation in Texas offers a dramatic contrast to the interventions fostered by citizens and environmental groups in the Pacific Northwest, New England, and California, the state's opportunities for energy efficiency and renewable energy development are probably now the largest of any state in the United States. A newly kindled interest in retail wheeling, competitive franchises, and a more market-oriented version of IRP planning could result in a new Texas energy boom, this one based on the development of clean power. A few features of Texas's IRP rule are noteworthy, particularly the requirement for creating local citizen advisory groups that can recommend competitive solicitations with portions earmarked for clean power.[7] Utilities are also instructed to develop a cost-effective portfolio that considers "nonfuel resources." Yet another provision that is attractive to renewable developers is that projects under 10 MW need no PUCT certification. This exemption could foster "direct access" green pricing programs.

Pat Wood, the young PUCT chairman, endorses green pricing as a tool to integrate renewable resources into the Texas energy mix. "The strengths of most renewable technologies lie in niche applications, such as distributed generation," he said, noting that he intended to "smoke utilities out on the issue of green pricing and get them to go ahead and see if it works." Wood acknowledged that such efforts to promote clean power "go beyond letting one sleep better at night." Under such programs, customers are "pre-paying capacity costs. Your variable costs will be zero. Under green pricing programs, rates will be less volatile. The benefit of wind and solar energy is a predictable price path," said Wood.

Texas Restructuring

According to Wood, Texas is looking to learn from California's mistakes. The Texas Legislature ordered an investigation into utility restructuring in 1995 when it approved the IRP law. Like the Pacific Northwest, the restructuring process is gathering input and developing principles *before* an actual proposal is on the table. "We are trying to re-orient the focus of IRP away from serving the vestiges of today's utility monopolies," said Wood in an interview, noting that PUCT did not want to "shock" the stock market with its deregulation plan. "We have the luxury of taking the best of both California and New England," he added, noting that on the issue of stranded cost recovery, easterners did a better job. "Sunk revenues deemed prudent will be recovered by utilities in Texas, but not O&M expenses or transmission and distribution expenditures," Wood stated. He argued that the fundamental political trade-off in restructuring was in granting utilities the recovery of stranded costs to allow customers more choices of power suppliers.

The positions of various parties on restructuring in Texas echo the debates elsewhere. Utilities, such as Houston Lighting & Power (HL&P), argue that utilities "are entitled to recover all invested capital in assets that have previously been deemed prudent."[8] The utility includes opportunity to earn a reasonable return on capital as one of its principal goals in restructuring. Not surprisingly, the Texas Industrial Energy Consumers disagree, arguing that customers should not be forced to pay above-market costs in any transition to a competitive market. Another critical issue in Texas is market power. The state's three biggest investor-owned utilities own as much as three-quarters of the power plants relied upon to meet demand.

Environmental and consumer groups have also put forward principles that echo what similar groups said in California: direct access opportunities should be granted to all classes of ratepayers at the same time; restructuring should not result in any degradation in environmental quality; and a renewable portfolio standard should establish minimum standards for a diverse mix of clean power sources. Public interest advocates agree with the industries on the issue of stranded costs, maintaining that above-market costs should be shared between utility shareholders and consumers. However, they are also willing to provide a few carrots to utilities, such as allowing a higher degree of stranded cost recovery for utilities that voluntarily divest generation from transmission and distribution functions.

A report commissioned by PUCT reveals how important stranded costs, particularly nuclear plants, are to opening up markets to clean power. The book values of nuclear power plants owned by four Texas utilities range from $2,700 to almost $4,400 per kW, compared to the $750 per kW estimate for a new combined-cycle power plant. These reactors represent a stranded investment of up to $15 Billion. That magnitude of uneconomic capacity on the system represents from 45 to over 100 percent of the equity of these utilities. The report concludes that the quicker the transition to market, the lower the amount of stranded costs and the greater the economic benefits for ratepayers.

Minnesota: Integrated Resource Planning and Nuclear Waste

The state of Minnesota offers some useful contrasts to Texas. It also offers important lessons for clean power advocates across the country.

Unlike Texas, Minnesota has a long legacy of strong environmental regulations. Like Wisconsin, the state's early Scandinavian immigrants fostered a culture where government regulation of private enterprise has been valued compared to the more free-wheeling cultures of California and Texas. It is also a state with a number of strong environmental laws and a tradition of support for non-hydro renewables such as wind power. Marcellus Jacobs, for example, designed one of the nation's most popular electricity generating wind machines in the late 1920s and eventually set up a manufacturing plant in Minneapolis that continued to manufacture these popular renewable energy systems until 1957. Many of the Jacobs' small but reliable wind machines are still in operation today.

Still, until recently, Minnesota relied principally on nuclear and coal to produce electricity. Several environmental groups like the Izaak Walton League have been advocating for years for more investment in energy efficiency and renewable energy resources. However, it was not until a nuclear issue became controversial that the state's largest utility began to seriously look at renewable energy sources as a viable supply option. Unlike other regions of the country, the nuclear controversy in Minnesota did not revolve around economics. Instead it was directly linked to the handling of nuclear waste.

Northern States Power (NSP) owns and operates two nuclear power plants located on a small island on the upper Mississippi River. The two Prairie Island reactors have a good operating history, but storage space for spent nuclear fuel was in short supply. In order to keep the reactors operating, NSP planned to build a nuclear waste storage facility next to the power plants where the spent fuel would be stored in dry casks. Those plans were challenged by the Mdewakanton Tribal Council, whose lands were next to the facility, and by local citizen activists.

In 1992, the state PUC granted NSP permission to construct an "interim" nuclear waste storage facility on Prairie Island. However, the Indian tribe challenged the decision in state court, arguing that state law required the state legislature to license any "long-term" nuclear waste facility. The law was a legacy of the state's fight to prevent the siting of a national nuclear waste repository in the state. Nuclear waste storage had been controversial in Minnesota ever since the federal government started looking for a permanent site to store the more than 30,000 tons of nuclear waste accumulating at America's nuclear reactor sites. In the early 1980s, twenty potential sites were identified; eight of them were in Minnesota.

A state court ruled in favor of the Tribal Council, arguing that since there was no permanent nuclear repository in the country, the facility to be located next to the Prairie Island plants should be considered long term, not an "interim" storage site. That ruling pushed the decision about licensing a spent fuel storage facility into the state legislature.

This court decision led to a bitter, drawn-out fight in the Minnesota legislature in 1994. In the end, the legislature, by one vote, approved enough storage capacity to allow the reactors to run another seven years, but only after the utility agreed to spend over $500 million dollars constructing several large-scale wind and biomass energy projects.

The compromise angered the Mdewakantons and citizen groups who were trying to close the Prairie Island reactors by limiting the amount of nuclear waste generated. The approved legislation granted NSP a total of nine storage casks. It also required the utility to go out to bid for 225 MW of wind power and 50 MW of biomass in 1998. Michael Noble, director of

Minnesotans for an Energy Efficient Economy, acknowledges the good and bad of this compromise. "Minnesotans are excited about the dynamic economic development opportunities associated with renewable energy. But they are equally, if not more, concerned about the ethics of producing nuclear waste and the craziness of storing this waste in the center of North America's largest watershed."

Minnesota has one of the nation's most aggressive IRP laws. The PUC has a "renewable default" policy that requires utilities to prove that renewable options are *not* in the public interest before using nonrenewable resources. The PUC also sets minimum energy efficiency standards for utilities. And while the law requires utilities such as NSP to invest 1.5 percent of their revenues on DSM, NSP proposed to spend 3.7 percent of its revenues—or $96.2 million—in 1994 and 1995. By the year 2008, NSP proposed to add 2,000 MW of new DSM.

A study sponsored by the Prairie Island Coalition Against Nuclear Storage released in 1994 projected that a mix of renewables and energy efficiency could enable Minnesota to retire not only the Prairie Island reactors but the 775-MW Monticello nuclear power plant and to meet all of the state's load growth over a 20-year planning horizon. If this program were enacted, the renewable share of the state's energy supply would increase from 5 percent to 42 percent by the year 2007. The state would also gain 122,000 construction jobs, 26,000 permanent O&M jobs, $73 million in state and local taxes, and household savings of $7,500 over a 20-year period.

Minnesota is not, however, ignoring moves to restructure the power industry. Advocates of renewable energy have joined hands with electric cooperatives, municipal utilities, and public interest groups representing the interests of low-income consumers to develop consensus principles for electric utility restructuring reform in Minnesota. These groups agreed to eight principles that they claim should form the basis for any future deregulation. Beyond Noble's group, members of the coalition are Clean Water Action, Cooperative Power, EnergyCENTS Coalition, Minnesota Municipal Utility Association, Minnesota Rural Electric Association, and the Izaak Walton League of America (Midwest Office).

The consensus principles state that wholesale competition should be allowed to take effect first, and its consequences should be thoroughly evaluated before the need for retail competition is considered. In addition, stranded costs must be subject to equitable recovery from departing customers. Environmental quality and stewardship of resources must be maintained through an effective integrated resource planning process or similar public process that assures consideration of cost-effective demand-side management and portfolio diversity, including cost-effective renewable resources.

While these groups have agreed to a common position on restructuring,[9] the controversy over nuclear waste storage on Prairie Island continues to simmer. In the 1996 legislative session, NSP tried to cut a deal with Native Americans to steer funds earmarked for renewable energy development into a resettlement program in exchange for support for greater nuclear waste storage capacity. Citizen activists were successful in killing this proposal by convincing a majority of legislators on the committee with jurisdiction over the issue to refuse any deal on waste storage engineered by NSP. Before the close of the legislative session, environmentalists proposed to use $8 million of the renewable funding to help relocate the tribes, if new accounting and disbursement provisions were set into place for management of the unused portion of the renewable energy fund.

Though this proposal was not approved, the debate highlights the kinds of social justice issues that can divide environmentalists and the other community groups such as Native Americans. In Minnesota, these groups and other citizen activists ended up uniting to quell plans for NSP to add nuclear waste storage capacity. They also came up with an alternative plan to add new renewable capacity beyond current state mandates while, for the first time, providing compensation to the community most directly impacted by the utility's nuclear plants. While this plan was also not approved in the closing days of the 1996 legislative session, it was an approach that undercut NSP's efforts to split citizen groups over the issue of radioactive waste.

The Role of the Federal Government

One of the undeniable trends in the emerging competitive marketplace for electricity is that the role of state governments in decisionmaking about new resources is being weakened. Instead, state regulatory agencies are looking to market participants such as utilities to make decisions about when to invest in new power plants and what kind of new resources to bring on-line. Both investor-owned and municipal utilities will have more discretion about power options in the future. SMUD chose to engage the local community in its decisions about new resources.

Still, SMUD's policy choices were influenced by federal and state regulations and laws. In the recent past, it has been state PUCs that have wielded the most clout in electricity matters. The stories of Minnesota, Texas, and New England illustrate this point. As market forces displace the need for state oversight, decisionmaking could filter further down to the community or individual customer level. However, the types of choices available to local communities and customers across the country could be limited by the actions of the federal government.

According to a number of observers, proposed federal policies could provide incentives for the oldest and dirtiest fossil fuel plants to run more often, a development that would wipe out gains advanced by local efforts to develop clean sources of power. Another danger is that states might develop a consensus on restructuring only to have federal regulators overrule that decision and undermine desirable community-based choices.

As markets are being structured to accommodate state and local needs, the FERC has begun to display a far more active role in regulation than was witnessed in the recent past. FERC's decision to nullify the California BRPU signaled the agency's newfound interest in managing affairs previously dominated by the states.

The underlying statutory authority for FERC lies with the Federal Power Act of 1935 as last amended by the Energy Policy Act of 1992. Under this authority, FERC decided to investigate how to open up the nation's grid to more wholesale competition. Two fundamental issues were addressed by FERC's Notice of Proposed Rulemaking, issued in March 1995, a document that would come to be referred to as the mega-NOPR.

The first issue was to order utilities to make transmission assets that they owned available to competitors on a "comparable basis" to their own use of the system. That meant they could not favor power plants that they owned in deciding how the transmission system should be used in wholesale power transactions. This "open access" policy was dictated by the Energy Policy Act of 1992, and there was widespread agreement that such a step was necessary to create a truly competitive market. States such as California, which were calling for the creation of an Independent System Operator, went beyond the FERC proposal in creating a "common carrier" transmission system.

The second major issue addressed by the FERC was stranded assets. The agency ruled that utilities would be eligible for 100 percent stranded cost recovery based on a "lost revenues" approach. Though this ruling only applied to stranded costs from wholesale transactions, some worried that this recovery method could be expanded to also apply to the much larger issue of stranded investments at the retail level. Indeed, Elizabeth Moler, FERC's chair, had called upon Congress to adopt legislation making FERC the backstop in the event state regulators do not provide utilities with 100 percent stranded cost recovery.

FERC has never been an institution that sought significant input from the general public. "FERC never went through an IRP phase," observed David Moskowitz with the Regulatory Assistance Project. "The institutional memory and focus of FERC is very different than state regulators." A joint filing prepared by CLF and the Center for Energy Efficiency and Renewable Technologies (CEERT) urged FERC to involve more public input by encouraging comprehensive regional "settlement agreements" like the effort underway in New England. That regional effort sought to link stranded cost recovery to the retirement of dirty power plants that could not meet a uniform environmental standard.

In addition, these environmental groups were joined by several utilities in calling upon FERC to prepare an analysis of the environmental impacts of its proposed open access transmission policy. This coalition charged that FERC's proposed transmission policy would increase the burning of coal to generate electricity and undercut regional efforts to meet federal air quality standards (see Figure 5-5).

According to an analysis prepared by the Washington, D.C.–based Center for Clean Air Policy, five of the nation's nine metropolitan areas designated as having "severe" ozone problems by the EPA could be impacted by FERC's proposal. These cities include Baltimore, Chicago, Milwaukee, New York, and Philadelphia which together have a population of about 60 million people. Increasing the use of older coal-fired capacity could increase

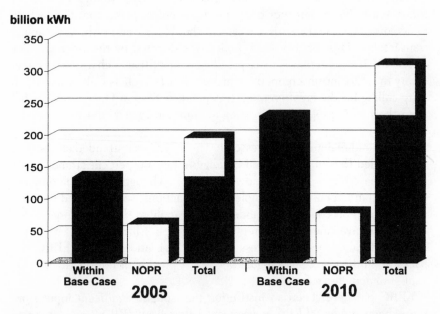

Figure 5-5: This bar graph illustrates projected increases in reliance on coal for generating electricity under FERC's proposal for open-access transmission without application of national uniform environmental standards. The increase associated with FERC's Notice of Proposed Rulemaking (NOPR) is an additional 50 million kilowatt hours in the year 2005. *Source:* Center for Clean Air Policy.

air pollution, particularly NO_x, a precursor to ozone, by 16,000 to 44,000 tons per year. Though these increases would occur at power plants in the Midwest, prevailing winds would carry a portion of this air pollution eastward. This raised concerns among utilities and states located along the Atlantic Coast. "Mitigating the adverse environmental impacts of open access, as we hope FERC will do, is clearly in the consumer's interest," commented James Crowe, executive vice president for United Illuminating, a Connecticut-based utility. "New Jersey and other northeastern states already must implement costly measures to reduce ozone health risks. It would be counterproductive for FERC to lower some electricity costs through open access, while, at the same time, increasing environmental burdens for New Jersey and the rest of the Northeast," stated Meredith Harlacher, senior vice president of Atlantic Electric.

State leaders also entered the fray on FERC's open transmission access policy. "We are simply seeking a level playing field," stated Republican Governor William Weld of Massachusetts. "A regulatory framework of open access transmission in conjunction with uniform environmental standards would remove the economic incentives for the use of dirtier plants," added Weld. He was joined by New Jersey Governor Christine Whitman who admonished FERC to "accept responsibility for the unintended environmental consequences of this rulemaking, and take a role in helping resolve the issue of ozone transport. The FERC rulemaking presents an ideal opportunity for federal and state governments to work together to manage successfully interdependent economic and environmental issues."

The approach advocated by a coalition of environmentalists, utilities, and state leaders was to shift costs away from Northeast industries that have already made substantial investments in air pollution controls and toward the owners of older and dirtier coal-fired power plants in the Midwest. This approach, they believed, could create incentives to replace old plants with cleaner sources of power.

Beyond its own rulemaking on transmission access, FERC will play a large role in the implementation of restructuring proposals that impact the wholesale power market, such as the CPUC's Power Exchange. Without the cooperation of FERC, the CPUC's plan cannot go forward. State PUCs such as California's have been the forum where many different constituencies have debated issues such as stranded costs and direct access and power pools. Some observers think that the federal government might even preempt the efforts of states to reform their electric service industry. James Caldwell, policy analyst for CEERT, argues that if parties are dissatisfied with a PUC order, they can still go to FERC, Congress, or the courts to block its implementation. "Electricity has always been a political exercise," he notes.

Congress, as well as state legislatures, could play a large role in the dereg-ulation debate. Efforts are already underway to have Congress shape poli-cies governing issues such as open access transmission and stranded costs. If Congress were to mandate 100 percent recovery of stranded costs, the op-portunities for states to develop alternative approaches to foster new invest-ment in cleaner technologies could be delayed for years. On the other hand, if concepts such as the renewable portfolio standard and systems benefit charge are enacted into federal law, a nationwide approach to energy effi-ciency and renewable energy sources could be strengthened.

Citizen activists can influence the development of the future structure of the electric utility industry at the local, state, and federal levels of govern-ment. Polls conducted in nearly every region of the country show that con-sumers prefer cleaner sources of power, even if they have to pay a little extra in the near term. If Congress or the state legislatures become the forum for articulating a broad framework for the reinvention of today's electric utility monopolies, then environmental standards should be part of a reform pack-age to replace dirty or inefficient power plants with new cleaner technolo-gies.

Notes

1. The region has the lowest electric rates in the nation. In 1993 the average system cost in the four Pacific Northwest states was 4.29 cents/kWh compared to the national average of 6.92 cents/kWh. The rate for the industrial class in the region was 3.22 cents/kWh; the U.S. average industrial rate is 4.86 cents/kWh.

2. BPA, without the knowledge or consent of the Department of Energy (DOE), en-tered into these contracts with DSI customers despite a DOE policy of not granting end-use consumers the equivalent of wholesale power purchase rights.

3. The Boulder, Colorado–based Land and Water Fund warns that any proposal to sell federal power marketing authorities such as the Western Area Power Administration (WAPA), which manages dams in California's Central Valley, to the private sector should recognize the wide variety of benefits these institutions provide. Any attempt to "de-fed-eralize" WAPA, the Southeastern Power Administration, and the Southwestern Power Administration should be viewed as an opportunity to promote sound energy and envi-ronmental practices. If such facilities, particularly WAPA as well as BPA, were to be sold in current market conditions, taxpayers would not be getting a good deal because current prices are so depressed. The Land and Water Fund argues that the price associated with any "de-federalization" scheme involving a private sector buyer should not fall below the net present value of repayment obligations and should reflect the true value of any trans-ferred assets. They also argue that transfer terms should include funding for mitigation of environmental impacts, including restoration of riparian and wetland habitats and a policy of using hydroelectric energy to firm up other intermittent renewable resources. Ideally, any sale would also take into account factors other than price, such as plans to protect the environment.

4. The total breakdown of BPA's costs show that 50 percent—just over $1 billion—are associated with generation. Transmission costs represent 18.7 percent, nonoperating nuclear expenses are 14.3 percent. The residential exchange—a program whereby investor-owned utilities are allowed access to cheap BPA hydro power—sits at 7.1 percent. Conservation expenditures and fish and wildlife were 6.5 percent and 3.9 percent of the 1994 total, respectively.

5. Interestingly, because of the lack of federal oversight, Texas has embraced a mandatory wheeling policy that requires a utility to transmit a seller's electricity. Wheeling deals have been going on in Texas since the early 1980s on a case-by-case basis.

6. This point is dramatized by the fact that in 1984 the average Californian used 267 kWh less than in 1977, while the average Texan used 1,424 kWh more. Overall, with 11 percent of the nation's population and 12 percent of national income, California consumes only 8 percent of total U.S. electricity. In contrast, with 7 percent of both national population and wealth, Texas consumes 9 percent of total U.S. electricity. Californians used less than half as much electricity per dollar of gross state product as did Texans.

7. Interestingly, when PUCT voted against the project, an advisory group insisted that citizens in the region truly wanted the project. PUCT then reversed course and allowed construction to proceed.

8. HL&P and Central and South West Corporation (CS&W) seem to prefer a pool-based system of restructuring. Texas Utilities, the other large investor-owned utility serving the large Texas electricity market, is more cautious, questioning the need to rush into any decisions until reforms proposed at the wholesale level are understood.

9. Local activists such as Michael Noble still favor IRP processes over voluntary "green pricing" programs to push renewable energy. "Why trade away IRP for a world where the only thing that matters is the price of electrons at the close of the day?" he asked. Noble's views underscore how approaches to clean power development in highly divergent states such as Minnesota and Texas need to account for the regulatory history of each state as well as their cultures.

Chapter 6

Last of the Monopolies?: The Future of the Electric Services Industry

"Everything has changed except our thinking."

—Albert Einstein

For the past three years, the debate about restructuring the electric services industry has hinged on the concept of "retail wheeling" or as it was later called "direct access." States throughout the country are investigating whether this approach, a wholesale power pool, or some other arrangement will result in sustainable lower costs for electric services (see Figure 6-1). It is an appealing idea to think that industrial, commercial, and residential customers could successfully negotiate lower prices for electricity if government and electric utility monopolies just got out of the way. Much of the political impetus for this argument has come from the biggest industries in the United States. Many of these companies—especially those involved in the manufacture or production of aluminum, steel, chemicals, paper, petroleum, rubber, and glass—are particularly energy intensive and are looking for ways to reduce costs as they compete in international markets.

Many of these energy-intensive industries have banded together to form the Electricity Consumers Resource Council (ELCON), the leading advocate for retail wheeling. ELCON is calling for major changes in federal law in order to completely deregulate power generation and to open up interstate electricity markets to end-use customers. To accomplish these goals would require the repeal of PURPA and the Public Utility Holding Com-

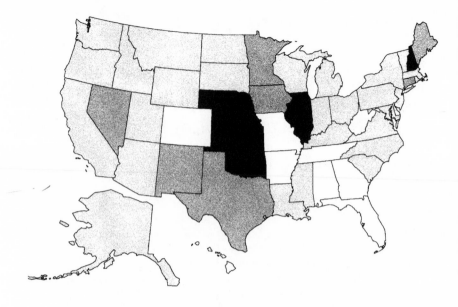

○ Utility Regulatory Commission Investigation

● Legislative Investigation

◉ Utility Regulatory Commission & Legislative Activity

Figure 6-1: Survey of states on utility restructuring issues, February 1996. *Source:* Hansen, McQuat, and Hamrin.

pany Act of 1935 as well as major revisions to the Federal Power Act. While ELCON's members account for only 4 percent of U.S. electricity consumption, their pleas have gained an audience in a Congress that is sympathetic to deregulation on ideological grounds.

These changes in federal law would stimulate the breakup of vertically integrated monopolies. While ELCON members believe that these changes will be to their benefit, they may be surprised to learn what has occurred in Britain since the privatization and breakup of the state electric utility monopoly. During the administration of Prime Minister Margaret Thatcher in 1989, the nationally owned electric utility was broken into local retail utilities, a national high-voltage grid, and deregulated power generation companies. The national government retained ownership of nuclear power plants since no private companies were willing to take on that responsibility. Any company is now permitted to build new power plants and sell electricity into the national power pool.

Since deregulation, electric rates in Britain have remained, on the average, roughly the same. However, prior to restructuring, large energy users such as chemical, cement, and aluminum firms received significant subsidies for electricity paid for by other consumer classes. Afterward, rates for these large industrial customers have gone up as hidden subsidies were removed.

The other change brought about by restructuring in Britain has been the closure of a significant number of older coal-fired power plants. Since the market for new power generation opened up, Britain has been gripped by what has been described as a "dash for gas" as a spate of companies have moved to build combined-cycle gas turbines. These new sources of power generation have forced the closure of 11,300 MW of coal capacity. These closures were triggered because older plants were inefficient and because coal became more expensive once subsidies for coal mining were terminated by the Thatcher administration. The end of these subsidies made natural gas much more competitive as a fuel for the generation of electricity.

These changes in the mix of the sources of electricity production have caused the British government to reconsider its position on global climate change. At the 1995 Berlin climate summit, Britain took the strongest position of any European country for reductions in carbon dioxide emissions. Also, the replacement of coal-fired power plants with more efficient natural gas–fired units allowed Britain to meet its commitments to reduce sulfur emissions without having to retrofit any existing power plants.

The British restructuring experience has served as a model for many countries in the developing world, Eastern Europe, and the former Soviet Union that had state-owned electric power monopolies. The motive for restructuring in these countries has usually been an interest by the national governments in stimulating foreign investment in electric generation. Particularly in Eastern Europe and the former Soviet Union, most power plants are notoriously inefficient and, in the case of nuclear power, often inherently unsafe. A strong case can be made that no country faces as severe a crisis in its power sector as Ukraine, which continues to operate two nuclear reactors at Chernobyl more than a decade after the catastrophic accident of April 1986. The president of Ukraine has, under pressure from Western nations, promised to close the two operating reactors by the year 2000.

Ukraine's economy revolves around energy-intensive industries such as heavy machinery production and metallurgy. Efforts to make these industries more efficient so Ukraine can compete in international markets bump into the issue of energy costs. Ukraine's power consumption is as much as eight times higher per unit of gross domestic product than the European average, mostly due to the inefficiency of electrical equipment in use. Under communist economic planning, the prices set for electric service bore little, if any, relationship to the cost of producing and delivering electricity. Because most consumers did not pay anything close to the full cost of service, they had little incentive to use electricity efficiently.

Since the collapse of communism, the government-owned electric utility monopolies in Ukraine have been broken up into four competitive generation companies, a national state-owned transmission system, and 27 local electric companies. Power will be exchanged by way of a wholesale power pool similar to the one established in the United Kingdom and recommended for California. The move toward market-based pricing for electricity will certainly cause discomfort to large energy users that had enjoyed government subsidies for electricity. Companies that want to become competitive will have to find ways to reduce energy costs. Given the inefficiency of Ukraine's electrical equipment, there is a potential gold mine of energy savings waiting to be uncovered. What is most needed in countries that are restructuring their electricity markets, including the Ukraine, are models for capturing efficiency savings on a cost-effective and self-sustaining basis.

Economic Efficiency or New Subsidies?

Of course, there are major differences between the structure of the electricity industry in the United States and that in counties that had nationalized power grids. What makes the United States system for delivering electricity so different from most other countries is the decentralized manner in which monopoly utilities were formed. It is therefore not surprising that restructuring would proceed in a very different manner in the United States than it has in Britain, Australia, or New Zealand, where public ownership has been predominant. Restructuring of the electric services industry in the United States, first of all, does not require the privatization of the ownership of power generation to create a competitive environment. There already are many owners of power generation waiting for state and federal regulators to create the new rules that will allow them to compete.

One troubling trend that is occurring parallel with the restructuring of America's electricity market is sharp increases in utility mergers, a trend that could reduce competition and frustrate efforts by citizens to have greater control over energy policy.

The pressures of competition affect utilities in different ways. Many utilities are positioning themselves either to take advantage of deregulated power markets or to protect themselves from competition. Those with low cost structures are eagerly looking at ways of capturing customers from neighboring utilities. Those with high cost structures, besides arguing for stranded cost recovery, are looking at strategies to lock in existing customers in preparation for a competitive environment. The most frequent measures now being used are special contracts, often negotiated in private, with large industrial customers. Obviously, large industrial customers have been encouraging this type of "competition."

However, ELCON's members have benefited from the *threat* of more competition among power generators. In many states, electric utilities have convinced regulators to allow them to cut special deals with large industrial customers to lessen their interest in finding other power suppliers. This has occurred in California with large cement companies; in Massachusetts with Raytheon, the large defense company; and in Michigan with the big three automobile companies. What these exclusive and sometimes undisclosed bilateral contracts have done is shift costs from the large energy users to other customer classes that do not have as much political or market leverage. While it is understandable for profit-seeking organizations to seek to shift costs to others, it certainly cannot be argued that this results in a more efficient electric services industry. The temptation to shift costs will continue as long as large power surpluses remain available in the wholesale power market.

However, if the restructuring debate remains focused on giving some retail customers access to the wholesale power market, the inevitable short-term result will be to shift costs from those who have access to those who do not. Unless, the *primary cause* of high electric costs is removed from the final rate equation, the electric power industry will continue to be inefficient and costly. This moves the discussion of restructuring back to the issue of "stranded costs." This odd piece of utility jargon really is a code word for nuclear power plants. There are, of course, some other utility "stranded assets." The fact remains, nevertheless, that if the nation's 109 nuclear power plants disappeared overnight, the restructuring of electric utilities would be relatively easy.

Some free market advocates argue that retail wheeling and the denial of full recovery of stranded costs will solve the problem of high rates and lead to a more competitive industry. They admit that there will be winners and losers in the short run, but in a reasonable period of time, costly power plants will be closed. The problem with this scenario is that during the transition competition for survival will force cost cutting at nuclear facilities that could compromise safety. The financial pressures can already be observed in some parts of the country.

At two nuclear plants in the Northeast, such cost cutting has already resulted in alleged safety violations, resulting in NRC investigations. In 1995 Northeast Utilities, based in Waterford, Connecticut, announced that it intends to reduce its nuclear work force of 3,000 employees by 700 to 1,000 by the year 2000. After a number of mishaps at one of the utility's nuclear plants (Millstone 1) were revealed in a major exposé in *Time* magazine, all four of the company's nuclear power plants were shut down by the NRC. Similarly, the Maine Yankee nuclear plant in Wiscasset, Maine, is being investigated after allegations that the company had misrepresented test results on the plant's cooling system.[1]

Growing numbers of industry analysts predict that the pressure to reduce operating costs at nuclear plants will force many to close in the near future. Though this outcome is now being acknowledged as a possibility, there has been almost no discussion about how a phaseout of nuclear power plants would be conducted in an orderly manner so that safety is not jeopardized during the transition.

Whither Nuclear Power?

Former NRC member Peter Bradford is one of the few people explicitly asking the question: Can nuclear power compete? His survey of industry opinion, and observations of what is happening at the nation's nuclear utilities, leads Bradford to the conclusion that it cannot. He then asks whether government should intervene to override this market reality. He makes the case for government intervention based on concerns for safety, reliability, and environmental impacts.

The compromise of safety at some nuclear power plants as economic competition intensified was predicted several years ago by former NRC Chairman Ivan Selin. He was concerned that competitive pressures would lead utility management to "cut corners," or reduce capital investments needed to keep critical plant equipment in top shape. Selin worried that some licensees would make decisions based on economic merits without adequate consideration of the risk consequences. The pressure to take cost-cutting steps will only become more fierce as utilities face the possibility of losing customers to competitors because of high rates.

Five nuclear power plants have been prematurely retired since the closure of Rancho Seco in 1989. In no case did the shutdown of these facilities jeopardize the reliability of electricity delivery. In fact, three nuclear plants have been closed on the West Coast, taking over 2,500 MW of capacity out of production while the price of power in the wholesale power market continues to decline. However, nuclear power accounts for only 16 percent of electricity production in California, Oregon, and Washington. It makes up 51 percent of power generated in New England and 39 percent in New York, Pennsylvania, and New Jersey. Although each of these regions has power surpluses allowing for elimination of some plant capacity, the reliability of electric supplies would be disrupted if all the areas' nuclear power plants were suddenly closed. Even a rapid phaseout might cause temporary power shortages resulting in price spikes before new supplies are introduced to balance supply and demand.

Subjecting nuclear power plants to economic competition would probably result in the replacement of many reactors over a relatively short period of time by natural gas–fired combined-cycle combustion turbines that have proven to be quick to build and efficient to run. This may be a reasonable way to lower the cost of producing electricity, but it would increase the amount of carbon dioxide emissions to a degree that is incompatible with commitments the United States has made as a signatory of the 1992 Convention on Global Climate Change. Over 160 countries have pledged to reduce carbon emissions, but the United States leads the world in producing these emissions. It is estimated that on an annual basis, the nation's nuclear power plants prevent the emission of about 133 million metric tons of carbon. If nuclear-generated electricity were to be fully replaced by fossil fuel sources, it would more than double the amount of carbon that the United States needs to reduce to meet its treaty obligation to stabilize greenhouse gas emissions at their 1990 level (see Table 6-1).

The nuclear power industry argues that it needs to be protected from competition in order for the U.S. government to meet its international global climate change commitments. Whether or not one agrees that all the environmental costs of nuclear power have been internalized, this argument needs to be taken seriously. However, it would be inconsistent to argue for government intervention to protect nuclear power while subjecting other truly carbon-free technologies to market forces. It makes no economic or environmental sense to subsidize the continued operation of nuclear power

Table 6-1: Global CO_2 Emissions (Millions of Tons)

Country	1965	1989	Growth Rate (%) 1965–1989	Share of World (%) 1965	Share of World (%) 1989
United States	948	1329	1.4	31.5	22.8
China	131	652	6.9	4.3	11.2
Japan	106	284	3.3	3.5	4.9
India	46	178	5.8	1.5	3.1
Germany	178	175	−0.1	5.9	3.0
Britain	171	155	−0.4	5.7	2.7
Rest of world	1432	3049	3.2	47.5	52.4
TOTAL	3012	5822		100.0	100.0

plants while foreclosing opportunities to develop renewable energy technologies or implement cost-effective energy efficiency measures that would allow for the closure of these facilities. If nuclear power is to be protected during the progression to a more competitive market structure for electric services, then there will be a need for government supervised IRP processes to assure that the transition is carried out in an orderly and least-cost manner.

Some highly respected energy analysts, notably Edward Kahn of Lawrence Berkeley Laboratory, argue that the transition to a competitive electricity market will require the nationalization of the U.S. nuclear industry. However rational this approach may appear, it seems highly unlikely that Congress would consider such a step at a time when budget deficit reduction and the devolution of federal responsibilities to states are such high national priorities. Nonetheless, the proposal points out the need for discussion of structural solutions to maintain nuclear safety and reliable operation during a transition period.

The *ad hoc* approach taken by the CPUC in the SONGS settlement does not adequately address this issue of a least-cost transition. That settlement, which has been touted as a precedent for dealing with nuclear stranded costs, allows the utility owners to continue to recover all upgrades made to the plant until 2003—whether or not those investments meet any least-cost standard. On one hand, it makes sense to eliminate disincentives for investments in nuclear safety. On the other, it is also reasonable to eliminate incentives to operate power plants when lower cost solutions are available. The CPUC approach virtually guarantees that SONGS will continue to be operated through 2005, regardless of any economic penalties to California's consumers that such a scenario may represent.

Now a fellow at the Regulatory Assistance Project, Peter Bradford claims recent utility investments in nuclear power plants such as SONGS should not be protected in moves toward competition. "Not a single dollar invested in an electric utility after 1980 came in justifiable ignorance of the possibility that it might not be recovered," he maintains. Bradford references the bankruptcies of WPPSS and Public Service of New Hampshire as adequate warning signs. Investors that bought stock at low prices are already being handsomely rewarded for their investments. Besides, returns on utility stocks have equaled or exceeded those available to other industrial investors.[2]

The biggest problem with stranded cost recovery is that the highest cost regions are those that have surplus capacity. These are also the regions—such as California—where there are incentives to delay, rather than embrace, competition. Bradford argues that the opportunity for collecting stranded costs should be tied to utility cooperation in providing open transmission access and support for social programs such as demand-side management. He observes, "Layers of environmental protection and resource planning depend on ratemaking practices that are threatened by competition just as investors are threatened." Bradford asks, "Can government really be more solicitous of (utility) stockholders than it is of the nation's lakes and lungs?"

Some environmentalists argue that nuclear power and energy from renewable energy sources should be forced to compete head-to-head in a "direct access" market. While this strategy has a certain appeal—since it allows consumers to express their choice for technologies in the marketplace—it is probably an unrealistic solution because of the inherent dangers associated with nuclear power plants operating in a competitive market. Nevertheless, continuing to subsidize nuclear power while slashing commitments to less risky and cleaner alternatives is not an appropriate policy choice either. A transition strategy that allows for the orderly phaseout of uneconomic nuclear reactors and replacement of this capacity with a mix of renewable energy projects and DSM measures can pave the way to a more competitive and sustainable electricity infrastructure that is more responsive to customer desires.

The Distributed Utility

Regardless of the method for arriving at a compromise for the buy-out of "stranded assets," there is already an undeniable trend toward "distributed generation." Rather than build transmission lines and distribution facilities to move electricity from central power plants to the consumer, some utilities like SMUD are looking at ways of producing power closer to the customer through many smaller power plants. With plants scattered throughout the utility system, and even located on customer premises, this arrangement can save money and minimize the need for rights of way through public and private property. It would also signal a return to a more decentralized electricity system similar to the one existing at the turn of the century.

Small generators like combustion turbines, wind turbines, fuel cells, and photovoltaics can be located near customers to provide power where it is needed, unclogging overloaded power lines, and deferring the need for upgrades in the distribution system. Distributed generation would be cost-effective even if these small power plants cost more per kilowatt than large, central power stations. Determining the cost-effectiveness of a distributed power plant requires a comparison not only with the cost of power from a central station but also with the cost of transmission and distribution equipment. Dispersed power sources are not hindered by the costs of rights of way, transmission equipment, or the electrical losses associated with pushing power through miles and miles of wire to reach a customer.

Some utility planners see a natural match between renewable technologies like PV and wind power and distributed power generation. In areas of the country that have large air conditioning requirements, feeder lines and electric transformers serving local businesses and residences are designed to meet maximum loads that may only occur for a few hours a year. Rather than replacing an operable transformer with a larger unit when an area grows, it may make sense to site a small power plant at a strategic location. That way the utility gets an additional increment of power and avoids an investment in the distribution system. An experiment testing this theory was conducted at a PG&E distribution substation located in California's Central Valley near the town of Kerman. Instead of upgrading the substation, PG&E installed a 500-kW PV power plant.

Another attractive technology for distributed generation is fuel cells. They are quiet and clean and can be sited in the basement of a large apartment house or hotel. All that is needed is a hookup to natural gas to provide the fuel cells with the hydrogen used to produce electricity chemically. The heat that is produced by a fuel cell could be captured for space heating or water heating in the building. The fuel cell might be installed by the utility with the heat being sold as an additional source of revenue. On the other hand, it might be installed by the building owner with excess electricity sold to the utility.

Quantifying the benefits of distributed generation is difficult. The PG&E study demonstrated a significant benefit for a transmission and distribution system in a rural area that was growing slowly. Calculation of the benefits of distributed generation is more complicated in densely populated urban areas with redundant feeder lines. Some utilities are skeptical of the concept because of the difficulty of measuring benefits. Furthermore, most low-voltage distribution systems have not been designed to operate with dispersed sources of generation. Control technologies need to be installed in the distribution system for safety and system protection. These control technologies are available, but many utilities do not have extensive experience using them. Some utilities are resisting consumer-controlled generation by levying large standby charges for back-up power or by making it difficult to connect to the distribution system. However, as competition for customers becomes more intense, utilities will likely try to accommodate customer preferences. Also, legislators in several states, including California, have enacted "net billing" laws that require utilities to buy back surplus power from small generators at distribution rates.

Most utilities do not have sophisticated methods to compare investments in transmission and distribution with alternative investments in distributed generation, load management, or energy efficiency. Transmission and distribution are often treated as fixed costs by utility planners, rather than costs that may vary based on decisions about power generation technologies and end-use efficiencies. Investments in these delivery systems are far from trivial. This equipment can represent more than half of a utility's new capital investment in some regions. Also, a significant amount of electricity is lost through the transmission and distribution system. Line losses are reported to account for about 6.2 percent of the electricity produced in the United States.

Carl Weinberg, former PG&E manager of R&D, compares most utility grids to irrigation systems—"delivering a commodity from a large reservoir to many customer sites." He argues that future distribution systems will more closely resemble computer networks, "with many sources, many consumers, continuous reevaluation of delivery priorities, and continuous management of faults." He believes that the system will be managed through two-way communication systems allowing customers and producers to interact based on priorities and costs.

One of the arguments used against the economics of distributed generation is that at the neighborhood level, load fluctuates considerably due to the decisions of individual customers to switch electrical equipment on and off. When loads are aggregated into large batches the fluctuations caused by individual decisions are smoothed over. The ability to bring together diverse loads was one of the reasons that larger power plants became economic and utilities became monopolies in the early days of the power industry. The closer a power plant is to the customer the faster it has to ramp up and ramp down depending on the customers' demand for power. Some new technologies like fuel cells, however, are very responsive to changes in load and cycle well—without a lot of wear and tear.

Distributed generation will become even more applicable when more powerful storage technologies become cost-effective. A particularly attractive technology is the flywheel battery—a technology that stores energy mechanically rather than chemically. This device works by accelerating a flywheel to high velocities when surplus power is available. The flywheel powers an electric generator when power is needed. Lightweight materials, advanced magnetic bearings, and near perfect vacuums will allow construction of flywheel batteries that can be accelerated to over 100,000 revolutions per minute, with coast down times that last for many months.

Amory Lovins envisions a distributed utility that is built with fuel cells and PV, flywheel batteries and ultracapacitors, all managed by smart controls "responsive to the real-time price signals that will permeate the grid." He foresees the time when households will have "beer-keg–sized" superfly-wheels buried in their backyards that will give people the opportunity to be neighborhood power brokers—buying and selling power in real time to meet their needs and those of their neighbors. Even more visionary is the idea of fuel cell powered cars that serve as power plants when they are parked, which is most of the time. Lovins notes that a U.S. vehicle fleet powered by fuel cells would have four times the generating capacity of all the power plants now connected to the grid.

Although the timing of the diffusion of these technologies is hard to predict, they will give communities and neighborhoods the tools with which they can reinvent electric utilities in ways far more profound than the modest restructuring proposals that are being considered today. Utilities that learn how to apply these technologies to increase customer value and give the customer control over energy consumption will be well positioned to thrive in a more competitive future. However, they may no longer be thought of as utilities. They will have become energy service companies.

Community Aggregation: Plugging into Neighborhoods

The debate over how soon the nation moves closer to the ideal of the distributed utility, with a greater emphasis on energy services that reduce and manage demand, is wide open. There is no doubt that the forces of competition can only accelerate a trend to smaller generation facilities, less waste, and fewer large investments in infrastructure.

Nevertheless, restructuring will ultimately be judged on how it meets the needs of the majority of consumers, including small businesses and residential customers. Customer choice needs to be a real option not only for the large industrial membership of ELCON, but for neighborhoods in cities such as Chicago, Knoxville, Salt Lake City, and Duluth.

As is the case in some other markets, small consumers may benefit from competition only if they can band together to purchase bulk power. After all, it was the benefits of load aggregation that led to the utility monopoly service territories in the first place. In a restructured industry, new ways of aggregating customers to allow for consumer choice will, no doubt, be required.

The first newly authorized retail wheeling experiments began in the service territories of Central Illinois Light and Illinois Power in April 1996. Though the program allows all customers within geographical areas to seek alternative suppliers, this initial foray into the realm of direct access limits residential aggregation to 500 customers. This low threshold, among other factors, will ensure that large industries will be the prime beneficiaries of the initial stage of this retail wheeling experiment.

A report issued in the spring of 1996 offers a peek into one model that has been dubbed "community aggregation." It is a strategy enabling local government to play an expanded role in assembling a portfolio of resources to meet the electric power needs of their citizens. Entitled *Can We Get There From Here? The Challenge of Restructuring the Electricity Market So That All Can Benefit,* the report was produced by the Boston-based Tellus Institute and the Madison, Wisconsin–based Wisconsin Energy Conservation Corporation. It addresses the reasons that two key features of the CPUC restructuring proposal—retail wheeling and real-time pricing (RTP)—are not enough to lower the state's high electricity costs for customers with relatively small electric power demands.

Though the CPUC has touted these options as offering consumers new choices, advocates for small consumers complain that there are still important obstacles to their participating in a market for power supplies. Most residential customers will not be able to reap any benefits if they act as individual customers. "It will be critical for small electricity customers to be aggregated into negotiating, bargaining, and planning groups, in order to overcome many of the transaction costs and market barriers they face. Our analysis shows that, absent aggregation, most small customers will see increased costs from RTP and retail wheeling—even if retail wheeling is assumed to reduce overall electricity costs by 10 percent," reads the report. Because profit-seeking aggregators are likely to focus their attention on large customers, "governmental entities with the specific objective of providing benefits to all small customers offer the greatest opportunity to promote equity under restructuring."

Under a proposal put forward by TURN, a local government agency could perform services similar to a private aggregator or broker, but instead of being motivated by profits, would provide the lowest cost and widest array of energy services in a specific geographic region. According to TURN lobbyist Lenny Goldberg, the proposal recognizes that "there are not going to be meaningful choices for residential customers unless they combine their consumer muscle. We view the creation of the franchise as the building block to allow residential customers to obtain the benefits of competition."

Experience from the natural gas and long-distance telecommunications markets suggests that 60 to 75 percent of small consumers will not participate in a competitive market for these services based on bill savings alone. Unless electricity providers are willing to offer products and services tailored to customers' special needs, markets for small customers will not be robust. At present, electricity providers are discouraged from moving into these potentially huge markets because high transaction costs do not warrant such risky ventures.

RTP only helps those customers with high electricity usage as pricing signals allow customers to shift significant loads to times of lower electricity pricing. Nevertheless, metering and equipment costs may not be justified since residential loads are relatively low. Aggregation, however, can produce substantial economics of scale because according to the report "there need only be one RTP meter and one marketing transaction for the entire aggregate. Even if marketing costs are higher for an aggregate, the benefit–cost balance will be much better on a per-customer basis, and net benefits can flow to all."

According to the report, there are three aggregation paths: formation of a traditional municipal utility; formation of a "municipal lite" utility that owns very little of the existing distribution system, but is eligible to pursue wholesale power purchases under the FERC open transmission rules; and formation of new local governmental entities to automatically aggregate customers within their geographical boundaries. The report concludes that the last option is the most promising. "Community aggregation by new local government aggregation entities may be the most equitable means of distributing system savings among customers. Such agencies should be authorized by state legislation and should be empowered to participate in wholesale power markets either alone or by participation in existing or newly formed joint action agencies."

Large industrial customers have argued that this franchise approach could inhibit the development of a more competitive electricity market. The report's authors disagree. They argue that if "geographic franchises can compete against each other, as well as brokers and sellers, then the electricity markets may become more competitive electricity services markets." Another concern is whether customer choice would be more limited under an exclusive franchise service model. A geographical franchise would preclude the "cherry-picking" of the most profitable customers in a region, but mandatory local government aggregation could diminish potential benefits of customer choice. To respond to these complaints, TURN revised its aggregation proposal to include an "opt out" clause to allow customers to purchase electricity service from generation companies or from aggregators or brokers. This flexibility is consistent with the notion of consumer choice, but it would likely maintain enough of a "quorum of customers" to make the franchise economic.

To ensure that new franchises adhere to public policy standards, the authors also suggest that the enabling legislation include a commitment to minimum standards for energy efficiency, renewable energy, and environmental factors that can be expanded by local decisions and clear standards for rate setting and service rules.

A different approach to bring the benefits of competition to customers, including nonresidential customers such as casinos, is an approach being pushed to transform the transmission and distribution system of Nevada Power into the Citizens Energy Services Cooperative, a proposal that would create the largest nonprofit energy co-op in America.

The brainchild of attorney Jon Wellinghoff and environmental and consumer groups, the proposal would allow Nevada Power to keep its generation facilities. However, the investor-owned utility's transmission and distribution facilities would be sold to the new co-op. According to Wellinghoff, the former consumer advocate for Nevada, the co-op would more efficiently distribute electricity to the utility's current 400,000 customers.

This approach to allowing consumers to have more say about their local power provider would result in tax and financing benefits of approximately 10 percent of Nevada Power's current gross annual revenues. According to the Land and Water Fund, these economic gains could be used to facilitate the breakup of the utility. These funds could also provide annual funding for investments in cleaner power sources. Though Nevada Power views the proposal with suspicion, this is one of the creative strategies that could transform current electric utilities into agents of progressive change.

Other utilities, such as Texas–New Mexico Power Company (TNMP), have also laid on the table new proposals designed to allow smaller customers to reap the benefits of any utility restructuring. TNMP calls its proposal "community choice," which would lead to the disaggregation of the current vertically integrated structure as well as encourage aggregation of residential demand. The plan would ultimately allow each community now served by TNMP to choose a menu of distribution services. Rates would then vary between communities, depending on levels of reliability and other factors.

Green Marketing

One does not have to look very far to see that "saving the environment" is being used as a sales pitch by many businesses to market a variety of products, from recycled paper to mutual funds. A large proportion of the population in the United States has demonstrated a willingness to pay more for goods that do not damage the environment. In fact, to help customers sort out the hype from the facts, environmental groups have organized a Green Seal program as a way of allowing businesses to demonstrate their commitment to the environment. Since electricity production can have a profound effect on the air we breathe, the water we drink, and land uses, it is not surprising that interest is growing in marketing "green power."

So far, only a handful of utilities have shown an interest in green marketing. However, as competitive forces transform the market for energy services, more and more utilities will be looking at green marketing as a way of providing a product that certain customers value highly. However, it will likely be non-utility electricity providers that devote the greatest attention to developing green markets. In areas where "direct access" is being implemented, "green pricing" has the potential to become a powerful tool to promote the development of new renewable energy projects. A number of renewable energy companies are beginning to team up with environmental groups to figure out how to market electricity that has been generated by wind, solar, biomass, or geothermal resources directly to customers.

Public opinion polling conducted since 1990 shows that U.S. consumers favor renewable energy sources and are "willing to pay" more for cleaner sources of electricity. At least 22 utilities have already conducted green pricing surveys to find out how much more consumers are willing to pay for renewable energy (see Appendix C). The results of these surveys are revealing. Nevada Power Company discovered that 18 percent of its customers would be willing to pay up to $10 more per month for clean power; 15 percent were willing to pay up to $25 per month extra. A survey conducted by Wisconsin Public Service Company shows that 1 percent of those responding to its survey would pay $40 per month plus a $250 installation fee for a rooftop PV system that saved the customer $27 per month, resulting in a net $13 monthly electricity expense.

Seven utilities—Arizona Public Service; Central Power & Light; the cities of Austin, Texas, and Tallahassee, Florida; Detroit Edison; Northern States Power; and Southern California Edison (SCE)—participated in a recent survey of residential customers to gather data about support for PV. This research showed that 69 percent of the customers polled stated that renewable energy was "very important." Only 3 percent said that these sources of power were "not at all important."

Four utilities have acquired renewable technologies through green pricing programs. Here is a brief breakdown of these programs, including technology, monthly surcharge, percentage of customers participating, and total amount of green kilowatts:

- Public Service of Colorado: small PV; $1.78/month; 1 percent customer base; 24 kW
- Sacramento Municipal Utility District: rooftop PV; $6/month; 0.01 percent customer base; 960 kW

- Detroit Edison: on-grid PV; $9.89/month; 0.01 percent of customer base; 29 kW
- Traverse City Light & Power: wind; $7.50/month; 3.1 percent of customer base; 600 kW

The utility programs are all fairly simple to implement. Green customers voluntarily agree to pay a premium for electricity, usually from 5 to 15 percent above the normal rate. With the additional funds, the utility acquires renewable energy sources. Often the utility will augment these "green funds" with rate-based funding, a technique SMUD used for its "PV Pioneer" program. SMUD's program is driven by policies established by its board, which adopted a sustained orderly development strategy for photovoltaics (see Appendix A).

While green pricing programs are only a small part of a utility's responsibilities, they promise to be a top priority for some of the new market entrants in a restructured electric service industry. Many renewable energy developers have seen potential new projects put on hold as state regulators debate changes in the industry structure. Many realize that in a surplus power market, regulators are unlikely to press for the development of new power supplies. In order to generate new business, these renewable energy developers will have to reach out to customers with innovative products. Non-utility providers can bundle products in a number of ways. They could sell electricity with a guaranteed minimum amount of generation coming from renewable sources. Or they could sell an all-renewable, super-green product. Alternatively, they could bundle electricity from renewable sources with energy management services to offer customers a fixed monthly price for servicing a building. The variety of green energy products will be limited only by the resources and the marketing ability of the organizations participating in the new electric service market.

Lessons Learned

The biggest challenge that electric utilities face today is to stop thinking like monopolies. That, however, is easier said than done. No utility has yet volunteered to give up guaranteed rate-based revenues in favor of market-driven prices. Few have invited competitors into their franchise areas to offer new products and services to their customers. Most have fought hard to get 100 percent reimbursement for stranded costs. Some have even argued that future costs at power plants, including regulatory commitments to improve environmental quality, should also be treated as stranded.

It is the rare utility that offers customers multiple service options. Many see downsizing as the way to prepare for competition, often eliminating the very programs that would allow them to differentiate their services in a competitive market. The programs being cut back include innovative energy management options, including DSM, and investments in cleaner, more decentralized technologies that offer diverse values to customers.

All too many utilities still believe that they are in a commodity business. For them, an electron is an electron. Indeed, there will be a commodity market for electricity. It will be organized primarily at the bulk power level. It will be a low-growth market with low profit margins. In most regions of the country, it will be a market dominated by power surpluses for years to come. The winners will be companies that move quickly to close obsolete, uneconomic, and dirty power plants and that squeeze more production out of existing economic power sources. Losers will be those who expect government regulators, or the courts, to protect them from the discipline of the wholesale power market.

While this scenario sounds bleak for many utilities, there is an up side. The retail energy services market will thrive and grow much more rapidly than the commodity market. A plethora of new market entrants will be pounding on doors, looking for new opportunities to serve the specific needs of customers. The energy services market will be very competitive, but it will also offer the opportunity for forward-looking entrepreneurs to earn substantial profits. The key to success in this open market will be to offer value-added services on both sides of the meter. Some energy service providers will offer customers hardware. Others will provide performance-based services. Some will bundle the electricity commodity with facility management services. Still others may manage multiple types of energy—not just electricity. No longer will customers look to a single source for meeting their energy needs. Marketing prowess will determine who is successful and who is not.

In a restructured electric service industry, prices will be set principally by the market, rather than by costs determined in complex and time-consuming regulatory proceedings. A key to success in this newly energized market will be the common denominator of customer service. Image will become increasingly important in building customer loyalty. Successful energy service organizations will aggressively identify new market niches and design special products to meet those markets.

Technology will be an important driver of the reinvented utility. Trends in electronics, information technologies, material sciences, renewable energy, and the mass production and miniaturization of hardware all point to distributed utilities as the wave of the future. It will be a future that is characterized by increased productivity, lower energy consumption, and less waste.

Successful energy service companies will stay on top of technology development curves and will institute ways to quickly introduce innovative infrastructure that can produce value and reduce costs.

The debate regarding creation of a new structure for the electric service industry will continue for some time. Even California is far from resolving all the issues that have surfaced. In many states the debate is just beginning. Most of the participants in this re-examination of the vertically integrated utility monopoly represent utility monopolies, independent power companies, large industrial customers, or other professional interests. They, naturally enough, are looking out for economic gains. They are also all familiar with the conventional regulatory process.

However, the issue of how electric service will be provided in the future is too important to be left for just a small group of insiders to solve. A larger set of constituencies—small businesses, residents, municipalities, rural communities, and urban neighborhoods—needs to be engaged in fashioning solutions from a community-wide perspective. The emergence of new forms of energy services and technologies can provide communities with the tools to have more control over electricity production and use.

General Principles for Restructuring

While the restructuring processes evolving in each state will depend on a variety of local factors and resources, some general principles to guide efforts to reinvent electric utilities can be discerned from the public debate that has taken place in several states over the past several years.

1. *Focus on the goals of restructuring first.* Perhaps the most important lesson learned in California is that there should be a debate on the goals of restructuring before developing specific policies. Different stakeholders will, of course, have different goals. It is therefore important to create a working environment in which all stakeholders—utilities, consumer advocates, large industrial customers, independent power producers, and environmentalists—can air their views and seek to find common ground.

2. *Protect public benefit programs during the transition to new market structures.* Among the misfortunes associated with the process used by the CPUC were severe cutbacks in utility energy efficiency programs due to the uncertainty about the new market structure. PG&E and SCE now say that these reductions were a mistake—both for their customers and for their future competitiveness. Nonetheless, much damage was done during the two and one-half years that the CPUC debated various restructuring models. This could have been avoided by assuring that public benefit programs would continue to be funded during a reasonable transition period by way of a systems benefit charge.

3. *Deal early on with the issue of stranded costs.* The highest priority issue for utilities is stranded costs. Utilities will argue that they should be given a reasonable opportunity to recover costs arising from past decisions. However, they should be expected to take steps to lessen such costs and the corresponding burden on ratepayers. Treating future cost obligations, including capital upgrades, as stranded costs will only delay a competitive market and create barriers for new market entrants. There are some other critical questions associated with stranded cost recovery: How long should the recovery period be? How likely is it that the estimate of stranded costs will change over time?

4. *Unbundle generation, transmission, and distribution costs so that competitive functions can be provided by multiple parties.* The many services currently required to produce and deliver electricity are bundled by the vertically integrated monopoly. In order for power generation to be truly deregulated it will be necessary to separate transmission and distribution costs from the cost of producing power. Unbundling generation, transmission, and distribution costs may require radical surgery—the splitting up of today's vertically integrated utility monopoly into new companies.

5. *Assure that all classes of customers have reasonable opportunities to benefit from restructuring.* There will be great temptations to shift costs between classes of customers as utilities are restructured and competition takes hold. Industrial customers have been the strongest advocates of customer choice and would likely benefit from a "retail wheeling" regime. Measures that allow residential and small business customers to be aggregated into buyer groups need to be developed as part of any comprehensive restructuring package.

6. *Develop policies that promote resource diversity in a competitive power generation market.* Advanced and renewable energy technologies, like solar, wind, and fuel cell power generation, can play a valuable role in promoting fuel diversity, managing risks, and reducing environmental impacts. In a market that gives priority to the short-term lowest cost resources, many of these technologies will be bypassed. If they are to play a larger role in the future resource mix, policies such as a renewable portfolio standard need to be devised to encourage the sustained, orderly development of clean power sources.

7. *Apply equivalent environmental standards to all power plants.* One of the big risks of retail competition is that it will provide incentives for the continued operation of old, inefficient, and dirty power plants and will deter investments in newer and cleaner technologies. One way to avoid this outcome it to require that older power plants meet the same environmental standards as new power plants.

8. *Assure nondiscriminatory access to transmission services.* A reliable electric grid requires that complex decisions about the use of transmission facilities be made on a short-term basis. Those decisions can have a major impact on which power plants are dispatched at different times. Responsibility for the control

and operation of the transmission system needs to be independent of vested financial economic interests in power plants.

9. *Encourage the development of distributed power generation.* Small-scale distributed generation located close to load centers offers multiple benefits for electricity consumers. Policies need to be put in place that at a minimum do not discriminate against the development of new resources like solar technology and fuel cells in the distribution system.

10. *Minimize incentives for distribution companies to promote the sale of electricity.* Local distribution utility profits should not be tied to the amount of electricity sales that takes place over the retail system. This policy is important to assure that investments in DSM resources are given the same consideration as new supply resources.

Just as important as these goals for restructuring is the path that is chosen to get there. Citizen input is critical to establishing a broad framework for a marketplace that responds to the desires of consumers.

Two Fundamental Questions for Citizen Groups

How should citizens judge restructuring plans being contemplated for their community or state?

Much of the debate over restructuring the electric services industry has been driven by a short-term focus on electric rates and a desire to improve economic efficiency. These issues are important, of course, because they influence regional economies and local employment opportunities. Yet these issues are not the only considerations policymakers need to address in creating a more consumer-oriented and competitive electricity marketplace.

SMUD has learned that the forces of competition can also challenge clean energy programs. Some attractive projects have been delayed, even terminated, due to unstable power markets that have affected the financial viability of clean energy developers.

Still, SMUD's experience provides a working model for citizens in other parts of the country. SMUD's success in cutting costs while promoting environmental progress has gathered international attention. SMUD phased out its most uneconomic source of energy—Rancho Seco—and replaced it with new investments in efficient gas-fired cogeneration, energy efficiency measures, and renewable technologies.

More recently, SMUD has had to make some changes in its policies in response to the cost pressures imposed by restructuring. SMUD continues to offer a wide variety of energy efficiency programs targeted to all customer sectors, but it has changed the delivery mechanisms to lessen the amount of capital expenditures and the near-term impact on rates. That has meant an emphasis on low-cost financing rather than direct rebates as the primary means of achieving energy efficiency savings from customers. It also has meant a shift in emphasis from encouraging the early retirement of inefficient equipment to making efficiency upgrades at the end of the equipment's useful life.

Because of the amount of surplus, low-cost electricity available in the wholesale spot power market, SMUD has deferred commitments to new power supplies that would increase the utility's capital debt. This strategy translates into a greater reliance on short-term power transactions and less reliance on building new utility-owned power plants.

Some of the changes initiated by SMUD have been driven by a concern for maintaining rates below those of alternative suppliers. However, that goal has been balanced against a longer term perspective that recognizes the value of resource diversity and improving the regional environment.[3]

Consumers in other parts of the country need to balance the interest in holding the line on the cost of producing electricity with other important economic, social, and environmental goals. In the final analysis, there are two basic questions that citizens need to ask of proposals to restructure electric utilities:

1. *Does the proposal promote sustainable social relationships?* One of the more remarkable achievements of the electric power system in the United States has been the creation of universal access to electricity at affordable costs. Of course, electricity came first to the banks, elite hotels, and newspaper publishing houses in downtown New York, Chicago, and Philadelphia. But within a generation it had been extended across America from the hills of Appalachia to forests of the Pacific Northwest. The fact that universal service was achieved was not an accident, it was the result of policies adopted by the federal and state governments. Policies like low-interest loans for rural electrification, lifeline rates for basic household needs, and low-income weatherization programs are all instruments of government rather than the market. They are important policies that have helped to bring the benefits of electricity to all Americans and have promoted social cohesion.

One of the issues that should come up in every state that is considering restructuring of its electric utilities is whether there will be an equitable sharing of the costs and benefits of restructuring. In most states, low-income programs have been developed to guarantee that the most vulnerable citizens have access to electricity. If governed solely by the criterion of economic efficiency, the interests of low-income customers will largely be forgotten in a deregulated market. In contrast, large users of electricity will enjoy advantages because of their increased leverage in the marketplace. Even well-to-do residential customers may not benefit from restructuring.

Aggregation of the demand for electricity of small business and residential customers will be necessary to obtain the economies of scale required to reap the savings of a competitive market. If creative approaches to customer aggregation are encouraged in the restructuring, then competition could benefit small business and residential customers and could promote sustainable social relationships between customer groups.

2. *Does it protect the environment for future generations?* No one questions that the generation of electricity has major consequences on the local and global environment. If driven primarily by a short-term focus on electric rates, deregulation of electric utilities will decrease incentives to invest in energy efficiency and renewable energy technologies. Without explicit support for clean power goals during a transition period, financial support will wither. Failure to build upon the technological advancements made over the last decade would damage environmental quality and increase long-term risks associated with burning fossil fuels. The answer lies in supporting policies like the public benefit charge and the renewable portfolio standard that have been proposed in several states to promote clean energy projects in a competitive market.

Responses to these questions and others can help frame the policy choices that each region faces in reforming today's electricity system. For over a century electricity has been at the center of modern civilization; it has truly revolutionized our way of life. It has allowed humankind to concentrate physical power in extraordinary ways. Now, it is being used to magnify knowledge and to instantaneously spread that knowledge to the farthest reaches of the planet. B. B. King's electric guitar, the Internet, and a host of gadgets that now define civilization all have been made possible because of electricity.

Electric utilities now face the greatest challenge since they came into existence about 100 years ago. The opening up of competitive power markets creates an opportunity for citizens to participate in the reinventing of electric utilities. SMUD's experience shows how one community linked competition with citizen action to create a cleaner electric service system. Though California's deregulation proposal was incubated under turbulent conditions, it helped develop key concepts that are now building blocks of reform in other states and nations. A wholesale power pool, an independent system operator, direct customer access, a public goods charge that can not be bypassed, and a renewable portfolio standard are now being debated across the country because of California's wide-ranging examination of electric utility restructuring.

The forces of competition are sweeping through the electric utility industry. How these historic changes affect each community and region will be determined not only by the marketplace but by local sources of vision and leadership. California, New England, the Pacific Northwest, Minnesota, and even Texas and the Tennessee Valley all offer useful lessons. Opportunities are opening up for citizens to participate in the transformation of one of the world's leading industries, one that has been a source of both wealth and pollution for more than 100 years. The new structure of the electric services industry will not only affect decisions about how we produce and use electricity in the future but will be critical in determining whether we will have a more sustainable way of life as we enter the 21st century.

Notes

1. In addition, Public Service Electric and Gas Company announced in early 1996 that the restart of Unit 1 of the Salem Nuclear Generating Station was being postponed indefinitely due to microscopic cracks in steam generator tubes. And Vermont Electric Cooperative, Inc., filed for bankruptcy in early April 1996, citing mounting concerns over its investment in high-cost nuclear reactors such as Seabrook and Millstone 3. A chief reason for the bankruptcy filing was a state ruling that the utility could only recover $22 million out of $60 million in nuclear debt from ratepayers.

2. A survey conducted in early 1996 showed that utility executive views on stranded cost do not show much confidence that the rest of the nation would follow California's lead. Only 44 percent of those responding to an annual survey conducted by the Washington International Energy Group believed they would be able to capture 100 percent of their stranded investment. The most surprising admission by these utility leaders was that not all utility costs were prudently incurred. More than three-quarters of the 400 that returned the survey mailed to 3,557 potential respondents acknowledged some liability. Still, 73 percent believed these costs should be recouped.

3. For those reasons SMUD has issued a request for proposal seeking to add 50 MW of renewable energy sources from 1997 through 2002.

Postscript

Update on the California Plan

In the last days of the 1996 session, the California legislature unanimously passed a comprehensive electricity market reform based largely on a restructuring framework proposed by Southern California Edison (SCE), large industrial customers, and the Independent Energy Producers described in Chapter 4. The legislation, put together after a series of intense late night negotiations, will now govern the largest electricity market in the United States.

The most notable new feature not previously contemplated in this compromise legislation was the creation of a state "rate reduction" bond authority charged with issuing debt instruments to buy down electric rates for residential and small business customers over a five-year period. Any debt issued by the authority will then be repaid over 10 years by these same customers. This populist twist to the final restructuring bill was added so legislators could claim they lowered rates for the majority of electric ratepayers in the near term; the measure gained over 70 coauthors (out of 120 legislators) during an election year.

Despite this rate reduction for small consumers, the biggest beneficiaries of the new law are California's investor-owned utilities. They fought hard to have a very liberal definition of "stranded costs" to be eligible for full recovery. Not only will SCE, Pacific Gas & Electric (PG&E), and San Diego Gas & Electric (SDG&E) be permitted to recoup all of their investments in uneconomic utility and nonutility power plants, but they will also be allowed to recover significant above-market operating and maintenance costs for their nuclear power plants. In 1998 alone SCE will receive more than $200 million above the expected market clearing price for the two nuclear reactors at the San Onofre Nuclear Generating Station (SONGS); PG&E will be allowed to charge more than $250 million more than what the market will offer for its Diablo Canyon reactors that same year.

The practical result of these nuclear operating subsidies will be to inhibit investments of new, lower cost and cleaner power plants at least until 2002. In effect, the restructuring law delays real competition for five years. Even though large customers and those choosing to purchase half of their electricity from renewable resources will have the right to shop for electricity bargains on the open market in 1998, the heavily loaded Competition Transition Charge (CTC) allegedly designed to merely recover sunk investments is estimated to total roughly 4 cents/kWh. The practical end result of this large surcharge is to make most transactions with alternative, nonutility suppliers very unattractive. (This is one reason that most large industrial customers are seeking to extend current discounted utility interruptible rates, subsidies that will exceed $200 million between 1998 and 2001.)

Though new clean power plants, including many renewable resources, could beat the price of California's existing nuclear reactors, customers choosing to buy directly form nonutility sources through direct access transactions would still have to pay for past mistakes represented by the CTC. This barrier to new investment is the single most important failure of California's final restructuring plan. A mistake that, if replicated elsewhere, will thwart the fundamental goals of any forward-looking reform of the existing monopoly-based electricity market.

Policies included in the final legislation addressing customer demands for clean power are a mixed bag. Exemptions to the CTC for those purchasing at least half of their power from renewables may be included in a subsequent fleshing out of the state's renewable energy policies, which could provide incentives for a direct access renewable energy market. But the renewable portfolio standard was rejected purely on political grounds. Senator Steve Peace (D-El Cajon), chairman of the legislative conference committee that produced the final restructuring bill, claimed it would raise rates in his district because SDG&E lagged far behind in development of renewable resources. Under the standard, SDG&E would have been required to meet the same level of fuel diversity as its larger investor-owned counterparts—SCE and PG&E. Such politics will likely play a role in other states' contemplation of restructuring proposals.

The final bill, nevertheless, includes modest levels of investment for clean power sources between 1998 and 2001: $872 million for energy efficiency programs for the state's three investor-owned utilities and up to $540 million to pay for the clean power premium associated with new and existing renewable resources. This funding will allow many clean power programs to continue. In addition, some $250 million was earmarked for research and development, though the bulk of these funds will be spent on transmission and distribution technology advancement.

Another important distinction between the final CPUC restructuring proposal and the 1996 legislation is integration of municipal utilities into commitments to clean power programs. Under terms of the bill, all of the state's public power agencies will be required to spend at least the same proportion of revenues on clean power and the R&D programs as the lowest spending investor-owned utility—a percentage estimated to be 2.4 percent of annual revenues. While this would not impact a progressive utility such as SMUD, since it already spends in excess of 5 percent of its annual revenues on clean power, it could boost commitments among smaller municipal utilities.

In the final analysis a few things are clear. Neither technological innovation nor the magic of the market will in and of itself reinvent the electric utility industry. The dynamic forces of politics, including a role for citizen activists, will be required to assure that increased competition leads to new investments in clean power. In the end, each state's debate over restructuring will be shaped by a variety of factors, including the influence of well-heeled special interests such as investor-owned utilities and large industrial customers. In California, those supporting clean power and social programs for the poor were divided, and though they won some important concessions, they failed to halt what will be one of the largest public bailouts in history.

The biggest drawback to the California plan is that it camouflages the failure of nuclear power. We hope that other state lawmakers and regulators do not follow in California's footsteps when it comes to shielding the current system's least desirable power plants from the forces of competition now sweeping the country.

Appendix A
Descriptions of SMUD Programs

Energy Efficiency Programs

Home Audits

A home audit program provides residential customers with a detailed technical analysis of their electricity use and recommends specific measures to reduce consumption. During the audit the customer receives compact fluorescent lamps, low flow shower heads, a hot water heater blanket, gaskets for electric outlets, and weather stripping and caulking at no direct cost. Since these measures are considered to be the equivalent of investments in new power supplies, all customers share in the costs of such installations.

At the time of the audit, the energy advisor writes up work orders for measures such as attic insulation, duct repair, and infiltration repairs. SMUD then arranges for contractor installation of the recommended measures. Finally, SMUD conducts a quality assurance inspection.

Home audits can reduce a homeowner's annual electric bill by $150 a year or more. A typical single-family home in Sacramento that is heated with electricity uses between $1,400 and $1,800 in electricity. During an audit of such a home an energy advisory might recommend replacement of an old, inefficient refrigerator with a high-efficiency model; suggest replacement of an electric resistance furnace and inefficient air conditioner with a high-efficiency heat pump; arrange an upgrade in attic insulation; and sign the household up for participation in the air conditioning load management program.

The following is an example of how the SMUD program works for a residential customer with electric space heating: The cost of a new refrigerator and heat pump is $4,250. SMUD gives the customer a $100 rebate for the refrigerator and a $400 rebate for the heat pump. The remaining $3,750 is financed at 8 percent interest. The average monthly savings on the electric bill is $48. During the four summer months SMUD provides a bill discount of $10 per month for participating in the load management program. Financing the loan costs the customer $35 a month. Thus, the customer savings are $23 a month during the summer and $13 a month for the rest of the year.

Energy-Efficient Refrigerators

Refrigerators are one of the biggest users of electricity in many homes. Starting in the mid-1980s major improvements were made in the efficiency of refrigerators. More efficient compressors were used and better insulation was added to the bodies of refrigerators. The energy consumption of new refrigerators is now less than half that of refrigerators manufactured 15 years ago.

Each year, about 7 percent of a utility's customers buy a new refrigerator. SMUD wants its customers to purchase efficient models even though they cost more up front. To do this, SMUD offers rebates to customers who purchase models that exceed the appliance standards by specific percentages. The more efficient the refrigerator is the greater the rebate. SMUD also dismantles the old refrigerators so that the refrigerant is recovered and salvageable materials are recycled.

The refrigerator program is popular with local appliance dealers who advertise SMUD's rebates on specific models. The dealers pick up the trade-in units when they deliver the new models and bring them to SMUD's recycling facility.

Solar Water Heating

It's hard to imagine a more wasteful way to heat water for household uses than by using electricity. It has been compared to slicing butter with a chain saw. A much simpler way to heat water to the temperatures that are needed for washing dishes and clothes and for bathing is to use the sun. Heating water for households using solar energy enjoyed a wave of interest in the late 1970s and early 1980s. Spurred on by tax credits and rising fossil fuel prices, dozens of companies jumped into this new market, particularly in California. But that early market experienced problems.

Some companies inflated the price of systems to increase the amount of the tax credits. Many of the solar water heating units used at the time failed after a few years. Installation practices were haphazard and many roofs were damaged. Sales practices were sometimes unethical. With the end of federal and state tax credits, the industry went into hibernation.

That false start was unfortunate since solar domestic hot water systems offer a cost-effective and environmentally benign alternative to electric water heating. For SMUD, solar water heaters are a valuable resource because they reduce summer peak demand.

In 1992 SMUD initiated a program to promote the replacement of electric water heaters with solar units. The program was carefully designed to avoid the types of problems that were prevalent in the 1980s. SMUD set strict standards for participation in its program, relying on a national certification system to ensure that units sold in Sacramento would be reliable. At the beginning of the program, SMUD employees inspected all units installed by certified solar contractors. Rebates and low-interest financing were tied to the performance of the systems.

Since 1992 over 3,500 solar hot water systems have been installed in Sacramento through SMUD's program. The average cost of a solar water heater in 1995 was $2,880. SMUD paid an average rebate of $820 and financed the remainder of the cost at 8 percent interest over 10 years. An evaluation of the program found that solar water heaters in Sacramento save on the average 2,465 kilowatt hours per year, or approximately 67 percent of the total household consumption of water.

Low-Income Outreach

Many low-income customers live in older houses with leaks and poor insulation. For these disadvantaged customers, SMUD set up a program to provide direct installation for weatherization measures and insulation. These items are installed under contract with community-based organizations that have credibility in these neighborhoods. The program also provides energy education workshops to targeted residential customer groups.

Equipment Efficiency Improvement

The purchase of big ticket electrical equipment, such as air conditioners or heat pumps, often occurs when an older unit fails or a when a home is remodeled. Usually a customer gets advice on the size and model of the new equipment from a vendor or installer. To encourage vendors to recommend more efficient units, SMUD provides rebates and low-interest loans.

Rebates are based on the efficiency of the equipment. For residential customers, 100 percent financing is available for central air conditioners and heat pumps. For commercial and industrial customers, rebates are provided for heat pumps and air conditioners, energy efficient motors and lighting measures. Low-interest financing is also available for small commercial customers.

Commercial/Industrial Retrofit

This program provides audits, as well as rebates and financing, to commercial, industrial, and agricultural customers. Schools and other public agencies are also included. To avoid lost opportunities, SMUD attempts to get all cost-effective measures installed for each audit conducted. The program targets lighting, air conditioning, heating, motors, refrigeration, and process items. Major projects have included several hospitals, large supermarkets, office buildings, and low-income apartment complexes.

Schools are also targeted for improvements in energy efficiency. A typical elementary school uses about 200,000 kilowatt hours of electricity each year. Improvements in lighting alone can save at least 40,000 kilowatt hours. This is worth about $3,000 per year. Such improvements cost approximately $6,000—about a two-year payback. Improvement in heating and air conditioning equipment can save another $2,000 per year.

New Construction

This program is designed to stimulate the design and construction of buildings that have greater efficiencies than is required by state building standards. Incentives that cover all or part of the higher costs of energy efficiency improvements are provided. Also, hook-up fees for for electric service are discounted for efficient buildings.

There are two ways to participate in the program, a standard path and a customized path. The standard path relies on incentives for individual measures such as insulation, windows, ductwork, air conditioning, and lighting. The customized path is performance-based and requires a detailed energy and cost analysis to evaluate energy savings. The customized approach takes into account the interaction of design features such as the reduction in air conditioning load from more efficient lights that give off less heat. It also provides a means to evaluate more complex design features such as passive solar design and thermal energy storage systems.

Demand-Side Management Programs, 1995

Program	Demand Savings (MW)	Energy Savings (gigawatt hours)	Cost ($1,000s)
Residential load management	5.5	n/a	2,688
Commercial load management	2.6	n/a	1,406
Water pump load management	0.4	n/a	141
Pool/spa load management	0.5	n/a	116
Residential conservation measures	1.4	8.1	4,519
Multi-family conservation measures	0.3	2.7	926
Shade trees	0.2	0.4	2,384
Solar water heating	0.4	2.7	1,408
Large commercial and industrial retrofit program	4.0	14.4	3,490
Small commercial and industrial retrofit program	3.8	19.5	4,580
Schools and public buildings	2.7	29.2	6,118
Energy-efficient refrigerators	2.1	15.4	2,730
Residential equipment improvements	2.0	4.3	2,587
Residential new construction	1.2	0.9	2,768
Commercial and industrial new construction	4.3	14.4	3,761
Commercial thermal energy storage	0.6	n/a	594

Cogeneration Projects

Carson Ice-Gen

SMUD's first cogeneration plant was placed into operation in September 1995. It is called "Carson Ice-Gen" and is located next to Sacramento County's wastewater treatment plant. The $143 million plant produces 100 megawatts of electricity and is powered by two advanced aeroderivative turbines manufactured by General Electric. One is used as a peaking plant; the other is connected to two steam host facilities and has a steam turbine that also makes use of excess steam.

Besides the ice company, the cogeneration facility provides steam to the county wastewater treatment facility which, in turn, uses the steam to digest sewage. The cogeneration plant also provides back-up power to the treatment facility so that in the event of a power outage, the county can continue to operate its facility and avoid discharging raw sewage into the Sacramento River.

Methane, which is a by-product of the treatment of wastewater, is tapped to help fuel the power plant. SMUD is monitoring the composition of the digester gas to see how it changes throughout the year. There is concern that contaminants could affect the power plant's catalyst that cleans up carbon monoxide. General Electric will also conduct periodic inspections of the turbine to determine how the digester gas affects the machine's durability.

Plant operation and maintenance is being handled by Carson Energy, the small company that developed the plant and brought Glacier Valley Ice to Sacramento. They developed the plant on a fixed-price basis. Once it became operational, it was turned over to the Central Valley Financing Authority, a joint power agency (JPA) set up by SMUD.

SMUD buys electricity at contracted prices from the JPA. The county and Glacier Valley Ice buy steam. Revenues from the sale of electricity and steam are used to pay off the debt on the project and to reimburse Carson Energy for the operation and maintenance of the facility. SMUD is responsible for supplying natural gas to the plant.

Procter & Gamble

SMUD's second cogeneration project, scheduled to be completed by mid-1997, is located at a Procter & Gamble manufacturing plant. It will be powered by two GE aeroderivative turbines connected to a 33-megawatt steam turbine. The total plant output will be 117 megawatts.

Procter & Gamble uses steam to convert coconut oil into a variety of products that are sold to other industrial customers and used by the company at other locations. The steam from the cogeneration plant is priced much lower than the steam produced by Procter & Gamble's own boilers. These savings induced Procter & Gamble to commit to keeping its Sacramento facility open. To assure the company of the reliability of the steam supply, SMUD agreed to significant penalties if it is unable to deliver sufficient steam.

The cogeneration plant will have a high thermal efficiency rating and will produce power at three cents per kilowatt hour. It is to be connected to a new gas pipeline that SMUD completed in early 1996.

Campbell Soup

SMUD's third cogeneration project will use a large industrial turbine rather than the aeroderivative machines used at Carson Ice-Gen and Procter & Gamble. Campbell Soup's needs for steam are almost three times those of Procter & Gamble, allowing for the siting of a larger combustion turbine.

SMUD selected a machine manufactured by Siemens because of its very low emissions. Siemens agreed to act as the developer of the project, a first for the German company in the United States. The project will produce 146 megawatts and is estimated to cost $181 million.

Campbell's corporate management came very close to deciding to close down its 47-year-old facility in Sacramento. If it had done so, over 2,000 jobs would have been lost. SMUD, by improving the economics of the large steam supply, helped convince Campbell to stay in Sacramento and renovate its production facility.

The cogeneration facility will replace existing boilers at Campbell Soup and result in a significant decrease in air pollution. Recycled water from the soup factory will be used in the cogeneration plant to create steam. The plant may be able to achieve the lowest level of nitrogen oxide (NO_x) emissions of any fossil fuel power plant in the world. With a catalyst added to the very clean turbine, SMUD expects emissions of NO_x to be 2.6 parts per million or less.

Siemens will operate the plant for the JPA, the Sacramento Power Authority. An operating contract provides incentives and penalties that depend upon the efficiency of the plant's operation. Because of Campbell's large steam requirement, and its variability, this is an especially important provision. Management of the steam flows into Campbell and the steam turbine will determine whether the operator earns a bonus or is penalized.

One of the major concerns about the Campbell project is the reliability of the Siemens combustion turbine. A blade failure at a power plant in Virginia

SMUD's Cogeneration Projects

Project	Capacity (MW)	Maximum Steam Flow (lbs/hr)	Capacity Cost ($/kw)[a]	Melded Cost (cents/kWh)[b]
Carson Ice-Gen	100	86,000	$1,202	3.3
Procter & Gamble	117	120,000	$1,177	3.1
Campbell Soup	146	320,000	$943	3.1

[a]Capacity cost is calculated in 1991 dollars.
[b]The melded cost is calculated at an 80 percent capacity factor in 1991 dollars.

in 1995 badly damaged the turbine and an electric generator. This raised questions about the performance of this new machine. The operations and maintenance agreement with Siemens, therefore, puts the company at risk for the turbine's performance. The company agreed to very favorable terms for SMUD because it wanted an opportunity to demonstrate to other potential buyers that the turbine was reliable. The Campbell Soup project is scheduled for operation in the spring of 1998.

Advanced and Renewable
Technologies Development Program

Solar Thermal

For over 20 years the U.S Department of Energy has supported the development of a large-scale solar thermal central receiver to convert solar energy into electricity. Central receiver systems use a field of mirrors called heliostats to track the sun and focus solar radiation at a receiver mounted on a tower. To be economic, central receiver systems will have to be 100 to 200 megawatts in size and must operate at capacity factors up to 60 percent. They can achieve this high level of output when thermal energy storage is incorporated into the system design.

An early design of a central receiver was built and tested in the early 1980s in the Mojave Desert. An advanced central receiver using molten salt for heat transfer and energy storage is now being prepared for commercial demonstration.

Commercialization of this technology is being sponsored by a consortium of electric utilities and the U.S. Department of Energy. The consortium will invest $39 million to convert an existing 10-megawatt facility in Barstow, California, from a water/steam system to one based on molten salt. Electric utilities providing service in Arizona, California, Idaho, Oregon, and Utah are supplying 50 percent of the funding for the project. The converted plant, called Solar Two, will operate through 1998. The goal of the project is to reduce the technical and economic risks associated with building the first commercial central receiver power plants.

Solar Two will be 100 percent solar fueled. Water consumption will be less than for fossil fueled plants. The salt is cheap, noncorrosive, and environmentally benign. Storage of the hot salt will make the plant fully dispatchable from noon to midnight. All components except the receiver have been tested in sizes and configurations for commercial operation.

A second solar thermal technology is the solar dish/heat engine system. This technology consists of parabolic dishes that track the sun with heat engines/generators mounted at the dishes' focal point. Parabolic solar dish systems are relatively small modular units that range from 5 to 50 kilowatts in size. They are limited in size because of the structural requirements of the movable dish.

Two types of dishes are being developed to concentrate the solar radiation for transfer to the heat engine. They are single-element and multifaceted dishes. Single-element dishes are constructed as a one-piece parabolically shaped dish. Multifaceted dishes are constructed of sections of nearly flat reflectors that are mounted in a dish structure to focus on the receiver.

Solar dish systems require efficient, reliable, and lightweight engines in order to realize their economic potential. Developmental work for dish systems has focused on the Stirling engine because of its high thermal efficiency, silent operation, potential for extended operation between overhauls, and multiple heat source capability.

The goal for the Stirling engine is to have it operate at least 50,000 hours between major overhauls. The most promising Stirling engine design uses a free piston with a linear alternator. The free piston design eliminates most of the mechanical bearings and allows the engine to be sealed. A linear alternator is built into the engine to produce electricity. This design has the potential for high reliability, long life, and low maintenance and production cost.

An important issue for solar dish systems is the heat transfer system. Advanced receivers are being developed to transfer the solar energy concentrated at the focal point of the dish to the Stirling engine. The advanced receiver is critical to the reliable operation of solar dish systems.

SMUD is leading efforts to form a consortium of utilities to provide a test bed and early market for solar dish/Sterling engines. Utilities from Florida, Nevada, Utah, Texas, and Oregon have joined the consortium.

Fuel Cells

Fuel cells are usually categorized by electrolyte type. Phosphoric acid, alkaline, molten carbonate, solid oxide, and proton exchange membrane fuel cells are available or under development. SMUD is currently supporting the commercialization of the phosphoric acid and molten carbonate technologies as well as following developments in the other technologies.

Fuel cell power plants have a number of characteristics that provide important benefits, including:

- High energy conversion efficiency (40 percent and higher) even at partial load.
- Responsiveness to load changes that makes them useful for peak loads, and they function efficiently at full or partial loads.
- Environmentally benign (very low emissions and low chemical pollution).
- Modular size and short construction lead times to enhance planning flexibility.
- Quiet operation that allows for siting in urban areas.
- Distributed siting, which lessens line losses and can defer transmission and distribution construction costs.
- Waste heat can be used in cogeneration applications.

For most fuel cells the hydrogen fuel is produced from natural gas, although other feedstocks can be used. For lower temperature fuel cells a reformer is required to obtain hydrogen from natural gas. For higher temperature fuels cells internal reforming is possible.

Phosphoric acid fuel cells are already commercially available and are being installed by utilities in Japan and the United States. Phosphoric acid fuel cells were developed by International Fuel Cells, which is a joint venture between United Technology and Toshiba. The units are completely packaged and can be delivered by truck. The 200-kW unit provides 760,000 BTUs per hour of thermal energy at 190°F.

When used in a cogeneration application, the overall energy efficiency of the unit is 82 percent. In contrast, cogeneration power plants convert 65 to 75 percent of the fuel they consume into usable energy. A cogeneration fuel cell will produce about half the carbon dioxide emissions of a traditional power plant and 99 percent fewer NO_x emissions.

Another fuel cell technology has been targeted for commercialization by a number of U.S. utilities. Although not yet available commercially this fuel cell uses molten carbonate as an electrolyte and has the potential to overcome some of the factors limiting the cost-effectiveness of the phosphoric acid fuel cell. Potential advantages of the molten carbonate fuel cell are:

- Greater fuel utilization (85 to 90 percent compared to 80 percent).
- Waste heat can be used to reform the hydrocarbon fuel without the need for supplementary heat.
- Less expensive catalysts (nickel rather than platinum).
- Waste heat is available at higher temperatures, opening up cogeneration potential in a number of commercial and industrial applications.

To promote the development of this technology SMUD has joined the Fuel Cell Commercialization Group, a consortium of 35 U.S. and Canadian electric and gas utilities that is funding the installation and testing of a utility scale molten carbonate fuel cell power plant in Santa Clara, California. The facility is rated at 2 megawatts and will have a fuel-to-electricity energy conversion efficiency of 54 percent.

If the demonstration unit is successful and if utilities commit to purchase 100 MW or more of early production units, the Energy Research Corporation has agreed to build a commercial-scale manufacturing facility capable of producing about 400 MW of molten carbonate fuel cells annually. The target price of early production units is $1,500 per kilowatt. Future units are projected to cost approximately $1,000 per kilowatt.

SMUD has established a working group to identify sites for molten carbonate fuel cells that offer the best combination of benefits. Potential sites include distribution substations and outlying distribution feeder lines; customers needing high-quality power, such as semiconductor manufacturers; small cogeneration situations, especially where high-temperature thermal output is beneficial; and areas near light-rail transit stations or other variable load centers.

The electrochemical process of the molten carbonate fuel cell is well suited for the gas mixture produced from biomass decomposition occurring in anaerobic digesters and landfills. Siting in conjunction with a biomass gasifier project is currently being evaluated by SMUD.

Fuel cells are a technology that can transform the economics of the power grid. They provide a set of benefits that are different from large-scale power plants. Utilities are still attempting to quantify the full spectrum of benefits and avoided costs that modular technologies like fuel cells can provide. Among the benefits they provide are high availability, lower reserve requirements, less reliance on peaking units, improved system reliability, power quality, faster return of capital, and ease of siting.

Biomass

Since the discovery of fire, biomass has been an important source of energy for the human race. It continues to be widely used in the developing part of the world but has been supplanted by fossil fuels in industrialized countries. Biomass fuels are quite diverse. They include urban wood waste, agricultural waste, and crops grown specifically to be harvested as biomass fuel. Biomass can be burned directly or gasified for later combustion or use in a fuel cell.

Some biomass resources are seen as attractive energy technologies because they also produce other important societal benefits such as waste disposal and pollution control. Using biomass products for fuel can also result in other valued by-products such as fertilizers.

In Sacramento, biomass applications are limited to the use of municipal and agricultural waste. Most generators of small waste streams have shown little interest in investing in biomass technology. Unless the technology can be supplied on a turnkey basis, growth in biomass applications will be small.

SMUD is currently investigating two local biomass applications. The first is to utilize gas collected from the Sacramento County landfill. The second is at the Folsom Prison Municipal Waste Recycling Facility. The Sacramento County landfill is the fourth largest landfill in the United States. It creates enough biogas to produce approximately 6 megawatts of power. Evaluations of the costs of gathering landfill gas and bringing it to a power plant have until recently been unacceptably expensive for electric utilities.

However, state law now requires that the county install a gas collection system for greenhouse gas control. Without an alternative use for the gas, it will be flared. Since installation of the gas collection system no longer needs to be the responsibility of the power developer, the project appears to be more economically viable.

Heating content of the methane component of the biogas is expected to be approximately 500 BTUs per cubic foot, about half that of natural gas. The gas can be used with only minor clean up in internal combustion engines. However, SMUD is examining the possibility of installing a fuel cell at the site. That makes an understanding of the gas composition more important.

Use of the gas in a molten carbonate or solid oxide fuel cell is attractive because it would maximize conversion efficiencies and avoid air pollution caused by burning in a boiler or internal combustion engine. SMUD is currently monitoring the production gas stream to determine the necessary gas clean-up procedures. If operation of a fuel cell is considered feasible, SMUD will evaluate installation in 1997.

SMUD is evaluating another application at Folsom Prison, where the Prison Industries Authority, a private firm, has contracted to separate and recycle garbage collected by the city of Folsom. Organic materials not suitable for recycling will be composted in anaerobic digesters. The composted bottoms will be sold for soil enrichment and the gases can be used to generate electricity.

The gases produced in composting are mainly methane, carbon dioxide, and water vapor. As in the case of the landfill, understanding the composition of the gas is important in considering whether a fuel cell can be used.

Fuel cells are already being developed for other markets. What is unique about SMUD's program is the combination of two technologies to produce energy in a modular power unit. SMUD is collaborating with the Department of Energy and the Western Area Power Administration on this project. The potential for the use of composted materials is high in California because nearly half of the mass at municipal solid waste facilities cannot be recycled.

Advanced Geothermal

Geothermal resources have been used for centuries as a source of heat. However, it wasn't until the 1950s that utilities began to figure out ways to tap heat generated by the earth to produce electricity. The most productive geothermal electric generation area in the world is located in California at an area called "The Geysers." SMUD has two power plants located in this area. However, over the last several years there has been a noticeable decline in the steam pressure in this area due to the installation of too many power plants.

Other geothermal resources have been identified, in California and elsewhere, where steam, hot water, or hot rocks could be used to produce electricity. Many of these resources tend to be lower in temperature and steam quality than those found at The Geysers. The lower temperatures require new methods to recover the energy contained in these heat sources in order to make electricity.

Power plants that do not directly use geothermal steam to power a turbine are called binary cycle power plants. This type of plant routes hot well water to a heat exchanger where another fluid is heated to its boiling point. The vapor that is boiled off can be used to drive a turbine to produce electricity.

One of the most attractive binary cycle technologies is called the Kalina Cycle, after its Russian inventor. It is expected to improve energy recovery by as much as 50 percent in lower temperature geothermal applications. It can also be used with other heat sources such as a solar thermal plant or the back-end of a combustion turbine.

The Kalina Cycle uses a mixture of water and ammonia as the working fluid. The temperature at which this mixture can be flashed into vapor can be adjusted by varying the mix of ammonia and water. SMUD, the California Energy Commission, and several other utilities have supported a 3-megawatt pilot project in Southern California.

Based on the success of the pilot project, commercial-sized plants are being proposed in and around SMUD's service territory. SMUD recently received an offer of the output from a commercial-scale Kalina Cycle power plant that is proposed to be built on the Steamboat Springs geothermal reservoir near Reno, Nevada.

The project would be about 13 megawatts in size and has been awarded Department of Energy funding contingent on a power sale contract. The prices offered by this project are very attractive and will be compared to other renewable resources that are to be identified through a competitive bidding process. This geothermal project would be considered a renewable resource because the hot brine would be returned entirely to the reservoirs and not released into the environment.

Evaporative Cooling

For many people evaporative cooling brings up images of "swamp coolers," which make a house feel like a sauna. But anyone who has ever sweated knows that evaporation is an effective way to cool down. And new developments in evaporative cooling are yielding systems that provide cooling with much lower humidity.

There are three types of newer evaporative cooling systems. They are indirect–direct, two-stage indirect, and hybrids. All three systems have the potential, with Sacramento's climate, to produce energy savings of up to 70 percent for commercial and residential applications. They can also improve indoor air quality by bringing in outside air.

The indirect–direct and two-stage indirect systems rely completely on evaporative cooling and use 100 percent outside air. Their ability to provide comfort during extremely hot and humid weather needs to be tested. If these systems prove to be capable of maintaining comfort during all summer weather conditions, they could replace conventional air conditioning in Sacramento.

Evaporative systems should be able to compete on a first-cost basis with traditional air conditioning and offer large life-cycle benefits to the customer. Indirect–direct systems will be the least expensive and hybrid systems the most costly. Indirect–direct systems are already commercially available at costs comparable to compressor-based cooling systems. They can lower peak demand and energy consumption by as much as 70 percent when properly maintained.

Hybrid systems combine evaporative cooling and mechanical cooling. These systems can be designed to provide adequate cooling under any conditions. Hybrid systems are expected to provide at least a 40 percent energy savings compared to conventional systems. However, because they require a compressor-based system they will be more expensive. From 1996 through 1998, SMUD will be installing evaporative cooling systems in 100 homes, one multifamily building, three commercial buildings, and three schools. These systems will provide demand savings of 676 kilowatts. This demonstration project is designed to create interest in evaporative cooling in Sacramento, to develop information on the cost of installing and maintaining these systems, and to foster a local infrastructure for the more widespread use of these technologies.

Geothermal Heat Pumps

Geothermal or "ground source" heat pumps are conventional heat pumps that use the earth as a heat source or heat sink. Geothermal heat pumps provide air conditioning in the summer and heating in the winter, just like air-to-air systems. But instead of exchanging heat with the outside air as a conventional heat pump does, these systems either extract or deposit heat into the ground.

During the air conditioning cycle the heat pump extracts heat from the house. The heat is transferred to a mixture of water and anti-freeze that flows through the heat pump. It is then pumped through a loop of pipes buried in the ground. The heat from the water is transferred to the cooler soil surrounding the pipes. The water then recirculates through the heat pump.

Geothermal heat pumps offer considerable efficiency improvements, particularly under extreme conditions. When the temperature of the ambient air in Sacramento is 110°F the temperature of the soil will be 55°F. However, because the earth is a poor conductor of heat, the soil surrounding the

SMUD's Advanced and Renewable Technologies Development Program

Technology Type	1996 Budget
Photovoltaics	$4,238,000
Fuel cells	$1,490,000
Solar thermal technologies	$176,000
Biomass	$213,000
Other advanced supply technologies	$389,000
Reflective coatings	$163,000
Evaporative cooling	$258,000
Geothermal heat pumps	$245,000
Other advanced consumer technologies	$126,000
TOTAL	$7,378,000

ground loop may rise to 85°F. Still, it is much cooler than the atmosphere and better able to absorb the heat coming out of the house.

There are two types of ground loops, horizontal and vertical. Horizontal loops require more space, while vertical systems require well drilling equipment at the time of installation.

Demonstration projects are being implemented in the residential and commercial sectors. For residential projects SMUD covers the cost of the ground loop. Houses of various sizes have been selected for participation. At least 25 homes will participate in the program from 1995 through 1997. Systems will also be installed for a low-income housing project and for three commercial buildings.

A typical geothermal heat pump installation on an all-electric home with 3 tons of air conditioning will provide a demand reduction of about 0.9 kilowatts and 3,200 kilowatt hours of energy savings per year. This will save the customer about $250 on their annual electric bill. SMUD intends to develop a working relationship with local developers to encourage the installation of geothermal heat pumps in all-electric subdivisions. The utility also expects to develop a commercial sector market for this technology.

SMUD's Steps for Stimulating the Development of a Sustainable Market for Photovoltaics

SMUD has adopted three key steps to stimulate the development of a sustainable market for photovoltaics. The first step calls for the early procurement of photovoltaics at prices above those required for conventional power generation technologies. Utilities such as SMUD can purchase relatively small quantities of PVs with minimal rate impacts. In 1996 SMUD budgeted $4.2 million to purchase 750 kilowatts of photovoltaics. The systems that SMUD is purchasing include 100 rooftop systems and several larger arrays. The utility also intends to purchase photovoltaics that are integrated into building materials such as shingles and curtain walls. Larger installations are being used as shading for parking lots or will be sited at distribution substations.

The second step is the sustained and orderly procurement of PV systems. A large, one-time purchase of PVs is likely to increase prices for new technologies, given the forces of supply and demand. To make the investments necessary to achieve reductions in the cost of PVs, manufacturers need to have substantial *and* sustained markets. SMUD's policy has been to increase the quantity of PVs purchased as the price per kilowatt declines. Since the program was put into place in 1993, SMUD's total costs for acquiring and installing PVs has declined by more than 20 percent.

The third step is to account for the costs of PVs over a commercialization life cycle rather than just over a project life cycle. The cost of a rooftop photovoltaic system is currently about $6,000 per kilowatt. At that cost they are not yet competitive with conventional generation technologies. SMUD, however, views a modest investment at this cost level as part of a strategy to boost cost-effectiveness over the long term. With a sustained, widespread effort on the part of SMUD and other utilities, prices are expected to decline steadily. If costs can be reduced to $2 per peak watt, then many cost-effective applications will be able to be identified within a utility system.

Electrical Vehicles

SMUD has been a pioneer in cleaning up air pollution from the transportation sector by encouraging the use of electric transportation in Sacramento. Ninety percent of the air pollution in Sacramento can be traced to fossil fueled transportation. Since 1990 the utility has promoted the use of light-rail, electric shuttles, and electric vehicles (EVs) for personal and fleet use. SMUD's efforts have focused on supporting advanced transportation technologies, stimulating local interest in electric transportation, and developing appropriate infrastructure for electric vehicles.

SMUD has promoted the introduction of practical, lightweight, state-of-the art electric vehicles. Over 100 electric vehicles are now part of SMUD's fleet. These vehicles are being used to test motor/controller and battery technologies. In addition, SMUD has worked with Sacramento's regional transit agency to acquire five electric buses that are used to shuttle passengers in downtown Sacramento.

The utility has also installed recharging stations at public parking lots, light-rail stations, work sites, and people's homes. Already, 29 EV charging stations are available with over 200 charging outlets. The number of recharging stations is expected to expand to over 200 by the year 2000. A special time-of-use rate has been adopted to encourage off-peak recharging. An Electric Vehicle Pioneer Network was created among community leaders and concerned citizens to support electric vehicle commercialization. The group assists in educating the public about opportunities for the use of electric vehicles and has supported local policies to increase electric vehicle use.

The city of Sacramento offers free parking for electric vehicle drivers in city-operated parking garages. The city and the county have adopted building codes that require all newly constructed homes to include wiring for electric vehicle charger installation. Also, the Sacramento Air Quality Management District offers up to $800 in incentives for electric vehicle purchasers.

SMUD also works cooperatively with automobile and battery manufacturers, finance companies, and state and federal agencies to promote the accelerated development of safe electric vehicles. SMUD is working closely with the U.S. Departments of Energy and Transportation and the Electric Transportation Coalition to develop a policy framework for "EV-Ready" communities. General Motors selected Sacramento for testing of its new electric vehicle, which it intends to bring to market at the end of 1996.

A total of 74 volunteers used General Motors "Impact" from February through July 1994. The average cost to charge the vehicle was 1 cent per mile. Approximately 90 percent of those who participated in the program said the electric vehicle met their transportation needs.

SMUD has joined with the Department of Defense's Advanced Research Projects Agency (ARPA) to support several electric vehicle initiatives. The vehicles are being developed at McClellan Air Force Base in Sacramento and are part of a national effort to demonstrate how military bases can be used to develop environmentally beneficial technologies.

Among the projects funded are development of an electric pickup truck prototype that has a range in excess of 100 miles. This vehicle uses advanced lead acid batteries, an advanced proton exchange fuel cell that can be used in a hybrid electric vehicle, and an electromechanical flywheel that can be used as a high-efficiency energy storage device and could replace batteries in future electric vehicles. As of 1996, SMUD ratepayers have contributed $3.46 million to the development of electric vehicles; federal and private sector funds have totaled $16.7 million.

Electric vehicles emit no tailpipe pollutants, an important consideration in highly polluted urban areas where poor air quality poses a health risk. For a utility with clean sources of electricity, like SMUD, electric vehicles are seen as an acceptable way of reducing urban smog. Using electricity can also significantly reduce the need for imported oil. Electricity needed to charge electric vehicles is abundant and is produced domestically.

Utilities are also interested in electric vehicles because of the potential to make use of idle power generation during off-peak hours. Batteries for electric vehicles could be charged during nighttime hours when power plants otherwise would not be used. This increase in the utilization of sunk investments can help to stabilize rates for all electricity consumers.

Appendix B

Promoting Renewable Energy in a Restructured Electricity Market

The debate over how best to implement a "renewables portfolio standard" in California has generated several competing proposals, all designed to create incentives for clean supply-side resource additions. As other states restructure electricity markets, many are looking to a renewable portfolio standard to integrate new, renewable energy projects into a competitive market for electricity.

Though the CPUC decision issued in late 1995 gave tacit approval to a "renewable portfolio standard"—a concept first put forward by the American Wind Energy Association (AWEA)—the transformation of this policy option into reality is not a simple process.

Under the AWEA model (see the diagram on the following page), which garnered the support of the geothermal and biomass industries, every retail seller of electricity would be required to include a minimum level of renewable energy in its supply portfolio. Technologies eligible to be included within the standard are wind, geothermal, solar electric, solid-fuel biomass, landfill gas, and existing solid waste-to-energy facilities. Projects that use 25 percent or less fossil fuel for firming would be granted full credit as a renewable facility. Projects that use more than 25 percent fossil-fuel backup would only be granted credit for the fraction of output generated by renewable fuels.

Beginning in 1997, a 0.4 percent per year increase in renewable capacity would need to be added to California's electricity system—a figure derived from the amount of renewable capacity that was to be added via the Biennial Resource Plan Update (BRPU). The standard would have to be met by

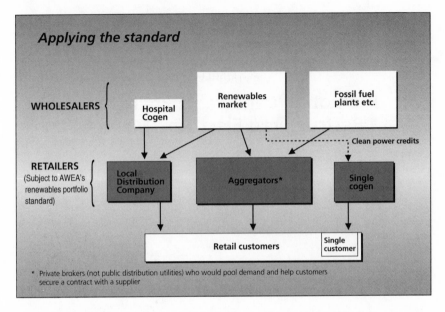

Applying the standard

AWEA's "Renewable Portfolio Standard." *Source:* Nancy Rader, *Windpower Monthly.*

retail sellers on an annual basis. If these sellers preferred not to become directly involved in renewable energy development, they could purchase renewable energy credits from other entities that co-develop or own existing renewable capacity.

The initial mandated amount of new renewable energy will be less than existing capacity—probably 90 percent of the 1993 renewable energy production figures—to ensure some competition to provide the lowest cost clean power supply to the state's consumers. Sellers that do not comply with the standard would be penalized. The mandate would also be repealed once renewable and market prices converge.

This fleshing out of the AWEA proposal in California raises some issues that other states need to consider in the development of a renewable portfolio standard:

• *What level of resource diversity will serve as the base for the "renewables portfolio standard?"* AWEA argued that 1993 should be the base year in California, since that is the high point for renewable resources on the state's system. Many states do not have significant non-hydro renewable resources on-line and therefore determination of a starting point may have to be based on other factors, perhaps linking future development goals to state air quality mitigation plans. Among the issues is the definition of renewables. Should utility-owned projects count? And at what rate should the standard's target escalate? AWEA suggested that California's standard start at about 11 percent of state electricity consumption, increasing to 13 percent by the year 2000. In Texas, AWEA proposed a goal of deriving 2 percent of the state's total electricity supply from renewables by 2002, a policy that would set aside one-quarter of the state's new capacity additions for clean, renewable sources.

- *Who should be subject to the standard?* AWEA has suggested that retail electricity providers be the ones governed by the standard, while the Independent Energy Producers (IEP) preferred imposing the renewables requirement only on regulated distribution utilities. This decision will be decided in political arenas. The advantage of the AWEA approach is that it ensures that all players will be complying with the standard. IEP argues that its approach is more realistic given the political reality that some currently unregulated market participants—large power marketers such as Enron—do not want to be responsible for complying with social program requirements.

- *How can technological diversity be promoted within the renewables standard?* AWEA did not want to create different bands for every technology within the renewable standard, because that approach would increase costs and limit new opportunities for wind power, the cheapest renewable energy option.

Vince Schwent, an energy specialist at the California Energy Commission and a member of a national network of utilities interested in distributed technologies such as photovoltaics, was concerned that technologies not yet cost-competitive would be frozen out of California's market. His proposal, therefore, creates several bands of technologies to ensure a sustained, orderly development of new renewable technologies.

Under Schwent's approach, 10 to 15 percent of the amount for renewables competing within the standard would be earmarked for bands of technologies other than wind, geothermal, and other technologies that are already close to competitive with traditional supply-side alternatives. For example, he proposed that 1.9 percent of the resources procured through the standard be earmarked for technologies in the 4.5 to 7.5 cents/kWh range—technologies such as biomass and solar thermal. The next tier would consist of 0.3 percent of the standard's resources for technologies costing 9 to 12 cents/kWh, the current cost of large-scale PVs. Schwent's final category—earmarked for 0.05 percent of the standard—would be set aside for technologies costing up to 30 cents/kWh. Schwent claimed that his tiered approach would ensure that not all of the new resources added would be wind and that the additional costs from these "small slivers of the market" would be well within "the noise" of market pricing.

A series of complicated issues surround the administration of renewable energy credits, including who will actually own these credits and be able to profit from their sale. The biggest decision in this policy realm is who should have the rights to credits generated by existing renewables under QF contracts—ratepayers, the QF, or the utility? The only justification for allowing utilities the rights to renewable credits from QF projects occurs in instances in which the utility is not entitled to recover the above-market costs through the CTC.

The chief rival to AWEA's proposal was put forward by IEP. However, infighting between the wind and renewable developer members of IEP and its gas-fired cogeneration and power marketer members prevented IEP from moving to the forefront of the political debate over the standard. Nevertheless, IEP believes that its approach to the standard—which would only be imposed upon distribution utilities—is the most viable politically since it would not require any new legislation. IEP believes that these regulated remnants of today's utility monopolies are the most logical entities to implement the standard since they will continue to have accounting, reporting, and metering relationships with all customers, including those choosing to become direct access purchasers of electricity.

Implementation of a renewables portfolio standard is now being considered on a national scale. Federal legislation incorporating the standard has recently been introduced by a Republican congressman from Colorado. In addition, the states of Massachusetts, Wisconsin, and Texas and the Pacific Northwest are considering some form of a renewables portfolio standard.

Among the other proposals that might complement a renewables portfolio standard is an idea forwarded by Rich Ferguson, national energy chair of the Sierra Club. He claims that imposition of a Competition Transition Charge (CTC), which is designed to collect from ratepayers stranded costs such as nuclear investments, could exceed 5 cents/kWh. He argues that this CTC should be waived for those customers who choose to purchase 100 percent of their electricity needs from renewable resources. "While there is considerable evidence that many customers would be willing to pay more for renewable electricity than they are paying now, there is no doubt that the CTC will be an enormous market barrier for most, if not all, renewable technologies," said Ferguson. "Customer incentives which offset the CTC would allow renewables to compete immediately against utility average costs—roughly 7.5 cents/kWh—rather than low utility marginal costs of 2.5 cents/kWh," he added.

All of these approaches, including proposals for a renewable portfolio standard, offer models to further the goal of sustained, orderly development of renewable energy sources. Proposals ultimately rejected in California should not be dismissed by other states. Each state market has unique attributes that may, or may not, be served by the policies adopted for California. The proposals cited will, no doubt, be offered to regulators, utilities, and advocates as possible avenues to meet the goal of adding cost-effective renewable power sources without the complicated regulatory processes of the past, perhaps best exemplified by the California BRPU.

Summary of National and Regional Surveys Affirming Consistent Public Support for Conservation and Renewable Energy

Compiled by the Renewable Northwest Project

Introduction

The Renewable Northwest Project (RNP) reports strong, consistent, and diverse support for the environment, and the continued acquisition of conservation and renewable energy resources, based on 13 national and regional surveys, plus information from utility focus groups all completed within the past two years. This support is broad-based, crossing utility customer classes, economic classes, political inclinations, educational achievement, geography, and ethnicity. The surveys and focus groups were performed by public and private utilities, public interest organizations, and industry associations.

This document summarizes key findings from these surveys and focus groups, and it includes the entity that commissioned the study, sample sizes, surveying dates, and when possible, the margin of error. Individual survey questions are quoted to provide some context.

National Survey Key Findings

Sustainable Energy Budget Coalition, Washington, DC
1,000 random registered voters surveyed December 1–10, 1995. ± 3.1% error.

• Over 70 percent of respondents believe global climate change is a serious problem. Seventy-five percent expressed willingness to pay higher utility rates if the money were used to supply electricity from renewable resources. A majority also wanted conservation and renewable energy resources to receive the highest priority for U.S. Department of Energy (USDOE) research and development funds.

Public Citizen, Austin, TX
400 random adult Texas residents surveyed September 29, October 2–3, 1994. ± 4.9% error.
- Seventy-five percent favored *requiring* nonpolluting electricity generation technologies to be responsible for one-quarter of new power supplies through the year 2005. Seventy percent were willing to spend up to $5 extra per month to receive electricity from renewable resources. When asked the same question, without a predetermined amount, 52 percent would still pay extra, some up to $15 per month.

Pacific Northwest Surveys and Focus Groups

Western Montana G & T, Missoula, MT
Results are weighted from surveys of member utilities taken in July 1995.
- Sixty-one percent agreed that their utility should only acquire resources having a positive or neutral effect on the environment. Seventy percent wanted the utility to promote programs to increase conservation. Forty-four percent wanted their utility to build, or purchase power from, wind and solar resources to meet loads five years into the future, if needed. These two renewable resources received the highest two voting totals.

Salem Electric Cooperative, Salem, OR
Responses to a June 1995 newsletter article sent to ratepayers.
- Seventy-eight percent of responses emphatically supported an investment in renewable energy resources. The supporters felt that all ratepayers, except for low-income people (150 percent of poverty level), should participate in the investment. The majority supported a rate increase between 4 and 8 percent to make the utility 20–40 percent "green."

Eugene Water & Electric Board, Eugene, OR
EWEB surveyed 400 customers, June 2–11, 1995.
- Customers prioritized conservation, wind, central solar, distributed solar, and geothermal resources as their top energy resource choices, in that order. Nearly 60 percent would spend between $1 and $12 extra per month on their utility bills to ensure the use of conservation and renewable energy resources for future energy needs.

Emerald People's Utility District
926 responses to 1,831 surveys mailed out in May 1995.
- Conservation, solar, wind, landfill methane recovery, and geothermal re-

sources, respectively, received the five highest levels of customer support for resource investment. Nearly 60 percent would pay extra on their monthly utility bill to support renewable energy resources; 50 percent would do so to fund conservation efforts.

Kenetech Windpower

801 telephone interviews with Oregon and Washington adults, 18 or older (401 in Washington; 400 in Oregon), May 23–25, 1995 by the Wirthlin Group.
• Eighty-six percent would replace lost hydro capacity with wind, even if their monthly utility bill increased by $9.

1,009 randomly dialed California adults 21 and older, surveyed January 20, 22–24, 1994 by the Wirthlin Group. ± 3.1% error.
• Ninety-five percent wanted a cleaner environment, and 82 percent supported the development of wind power. Given the option of personally awarding energy supply contracts, 68 percent would choose wind power.

Sacramento Municipal Utility District, Sacramento, CA

401 residential and 392 business customers surveyed in February 1995. Residential customers were selected from all single and multifamily households within the service territory; and business customers were chosen from agricultural and commercial customers with less than 1-MW billing demand.
• More than 80 percent of residential and business participants selected environmental responsibility as an important utility service; nearly 75 percent in both customer classes selected "promoting renewable electricity production" as an important service. Additionally, 43 percent of residential and 38 percent of business customers would pay 5 percent extra in their monthly bill to promote the use of renewable energy resources.

Portland General Electric

766 random phone interviews (400 residential, 366 commercial), July 1994.
• Sixty-one percent of residential customers and 65 percent of business customers wanted PGE to prioritize development of renewable energy resources (solar, wind, geothermal specifically) in the next five years.

59 focus group participants (27 residential, 32 commercial), selected by annual electricity consumption (commercial) and pre-tax household income (residential), February 28 to March 2, 1994.
• Nearly 100 percent of residential and commercial customers felt that it was important for their utility to be environmentally responsible. Renewable resources filled 70 of 100 megawatts in a sample resource mix as chosen by participants. Ninety percent of residential and commercial customers felt that generating electricity from renewable resources was important. Over 90 percent of residential *and* commercial customers felt that energy efficiency programs were important.

Washington Water & Power
300 randomly chosen customers surveyed July 27 to August 1, 1994. ± 5.7% error.
• Ninety-five percent felt that WWP should continue to offer energy efficiency programs, even if WWP had enough energy to meet customer loads into the foreseeable future. Eighty-three percent agreed that if new energy efficiency programs were available, and funded by a $1 per month surcharge on all customer bills, then WWP should offer such programs. Fifty-seven percent wanted WWP to offer energy efficiency programs regardless of whether customer bills were lowered.

Snohomish County Public Utility District
Approximately twenty-five people in focus groups completed August 1995.
• Customers expressed support for spending up to 10 percent extra per month on their utility bill for renewable resources.

Conclusion

The message presented within these results is clear: the American public in general, and particularly those in the Pacific Northwest, feel that environmental preservation is important and that their utility should prioritize conservation and renewable resources. Many are willing to pay higher rates if that led to use of these resources.

National Survey Sample Questions and Answers

The following are the full text of selected questions and answers from the listed surveys and focus groups, providing documentation for the conclusions drawn above.

Sustainable Energy Budget Coalition (December 1995 survey)
• "In your mind, how serious a threat do you think global climate change, also known as global warming, caused by emissions from the combustion of oil, gasoline, and coal is?"

Very Serious	35.5%	
Somewhat Serious	35.4%	70.9% think it's serious.
Not Too Serious	16.0%	
Not a Threat at All	8.7%	
Don't Know / Refused	4.4%	

• "Which of these research and development programs do you think should receive the highest priority for funding in the [federal] Department of Energy's research and development budget?"

R & D Programs	% in Favor	
Renewable Energy Involving Solar, Wind, Geothermal, Biomass, and Hydroelectric Power	34.1	
Technologies to Improve Energy		55% prefer renewables & conservation.
Efficiency and Conservation	21.0	
Natural Gas	9.0	
Fossil Fuels Such as Oil, Gasoline, and Coal	8.6	
Nuclear Power	8.5	
None of These	4.8	
Don't Know / Refused	14.0	

• (Only those who believed that the United States should reduce oil imports were asked this question—in total, 75 percent of those surveyed.)

"A number of options have been suggested for reducing oil imports. As I read each of them, please tell me whether you favor or oppose the proposed policy. And would that be strongly (favor/oppose) or just somewhat (favor/oppose)?"

Support the development of renewable energy alternatives to oil.

Strongly Favor	66.4%	
Somewhat Favor	23.9%	90% favor renewables.
Somewhat Oppose	4.1%	
Strongly Oppose	2.1%	

• "Suppose you have the chance to choose your electric company the same way you now can choose your long-distance telephone company and the choice were between a utility company which uses coal to generate electricity and a utility company that produces electricity using cleaner, but slightly more expensive renewable energy sources. Of the following, which indicates how much more you are willing to spend per month for electricity generated from cleaner renewable sources":

Up to 2% per month	23.2%	
Up to 5%	25.6%	
Up to 10%	19.2%	75% willing to pay more.
Up to 20%	4.8%	
More than 20%	2.5%	
I wouldn't pay any more for electricity generated by renewable sources	23.9%	
Don't Know / Refused	0.8%	

Pacific Northwest Survey and Focus Group
Sample Questions and Answers

Western Montana G & T (July 1995 surveys)

• "Suppose that your electric utility needs to build or purchase a new generating facility to meet customer needs 5 years from now. Which of the following would you prefer that your utility build or purchase?"

Windmills	23%
Solar Plant	21%
Hydroelectric Dam	19%
Natural Gas Combustion Turbine	13%
Wood/Municipal Waste Burning Plant	10%
Nuclear Plant	5%
Coal Plant	3%

• "Next I'm going to read some statements. Please use a scale of 1 to 5, where 5 indicates that you strongly agree; 3 indicates that you are neutral, and 1 indicates that you strongly disagree. How would you rate your agreement with the statements?"

"The utility should promote programs that increase the conservation of electricity."
AGREE 70%; NEUTRAL 25%; DISAGREE 5%.

"The utility should only acquire resources that have positive or neutral effects on the environment."
AGREE 61%; NEUTRAL 32%; DISAGREE 7%.

• "Do you feel that it is more important for your utility to promote programs that increase the efficient use of electricity, or to keep rates as low as possible?"

Promote efficient use of electricity	53%.
Keep rates as low as possible	51%.

Eugene Water & Electric Board (June 1995 surveys)

• "Assuming similar costs for each resource, how strongly do you support or oppose EWEB's involvement in each of the following on a scale of 1 to 5, where 1 is STRONGLY OPPOSE and 5 is STRONGLY SUPPORT?"

Resource	Average Score
Conservation	4.7
Wind	4.2
Central Solar	4.1
Individual Solar	4.1
Geothermal	4.0

Hydroelectric	4.0
Methane (Landfill Recovery)	4.0
Cogeneration	3.9
Natural Gas	3.5
Coal	1.8
Nuclear	1.7

Emerald People's Utility District (May 1995 survey)

• "How much more would you be willing to pay in your *monthly* utility bill to ensure that *renewable resources* (solar, wind, geothermal, etc.) are used to meet our future energy needs?"

Amount	% Responding	
Nothing	41.1	
$1.00	33.5	
$5.00	16.7	59% would pay more.
$7.50	4.4	
Other	4.3	

• "How do you feel about the resource shown, as well as the following other energy resources?"

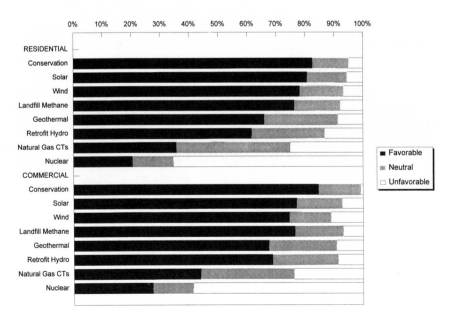

Emerald PUD residential and commercial customer energy resource favorability rating.

Portland General Electric (February–March 1994 focus groups)

• "Let's take a look at the resources PGE currently uses to acquire electricity. What do you think about these resources?" The following general responses were recorded:

"A higher than expected percentage of purchased power."
"Disappointment in the lack of non-hydro renewables."
"The smallness of the energy efficiency contribution."

• "Let's imagine that each of you is a resources manager for PGE. As a resource manager, you have to acquire 100 new megawatts of electricity. Decide which resources you would choose to acquire these 100 megawatts." (*If asked about cost of individual resources, let participants know that cost is not a variable they need to consider.*)

	Average Megawatts		
Resources	Residential	Commercial	
Green Resources (net)			
Energy Efficiency	17	14	
Solar Power Plant	8	9	
Geothermal	10	5	
Photovoltaics	9	5	54 and 46 MW of
Wind	8	5	non-hydro renew-
Biomass	2	8	ables for residential
Hydro	16	17	and commercial,
Green Resources Subtotal	70	63	respectively
Gas-Fired Cogeneration	5	10	
Gas-Fired Turbine	12	8	
Purchased Power	5	9	
Coal	1	3	
Nuclear	3	1	

Maps Depicting States with Commitments to IRP, DSM, and Renewable Energy

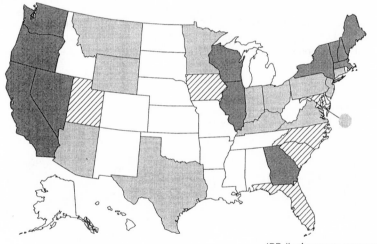

IRP (lacks many components)

IRP (lacks one component) ⊘

IRP (adopted and implemented) ●

States whose utilities have an IRP framework established. *Source:* Jan Hamrin and Nancy Rader. 1993. *Investing in the Future: A Regulator's Guide to Renewables.* Washington, DC: National Association of Regulatory Utility Commissioners.

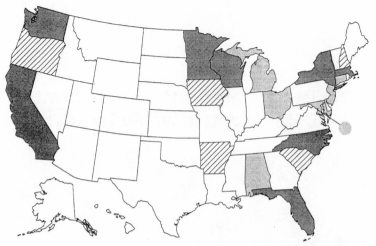

utilities with largest DSM expenditures as a %age of retail revenues ⊘

utilities with largest DSM expenditures ◌

states with utilities having both of the above ●

States whose utilities have large DSM expenditures. *Source:* Jan Hamrin and Nancy Rader. 1993. *Investing in the Future: A Regulator's Guide to Renewables.* Washington, DC: National Association of Regulatory Utility Commissioners.

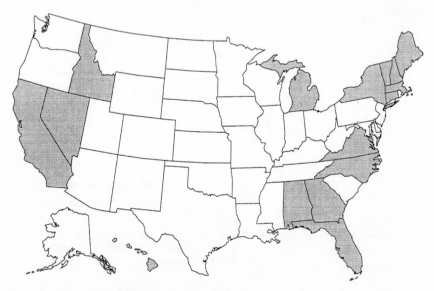

States leading in renewable resource capacity. *Source:* Jan Hamrin and Nancy Rader. 1993. *Investing in the Future: A Regulator's Guide to Renewables.* Washington, DC: National Association of Regulatory Utility Commissioners.

Glossary of Energy Terms

Active Solar Energy   Solar radiation used by special equipment to provide space heating, hot water, or electricity.

Avoided Costs   A regulatory term for the amount of money that an electric utility would need to spend for the next increment of electric generation that it instead buys from a cogenerator or independent power producer. Federal law establishes guidelines for determining how much a qualifying facility (QF) gets paid for power sold to a utility.

Base Load   The lowest level of power needs during a season or year.

Base Load Power Plant   A power generating facility that is intended to run constantly near its full capacity.

Biomass   Energy resources derived from organic material. These include wood, agricultural waste, and other material that can be burned to produce heat energy. They may also include algae, sewage, and other organic substances that can be used to make energy through chemical processes.

Blackout   A power loss affecting many electricity consumers over a large geographical area for a significant period of time.

Brownout   A controlled power reduction in which the utility decreases the voltage on the power lines so customers receive weaker electric current. Brownouts can be used if total power demand exceeds the maximum available capacity. The typical household does not notice the difference.

Bulk Power   Large amounts of electrical energy transmitted at high voltages.

California Environmental Quality Act (CEQA)   Enacted in 1970 it established state policy to maintain a high-quality environment and set up regulations to inhibit environmental degradation.

209

Capacity The maximum amount of electricity that a power plant can produce under specified conditions. Capacity is measured in megawatts and is also referred to as the nameplate rating.

Capacity Factor A percentage that specifies how much of a power plant's capacity is used over time. For example, typical plant capacity factors range as high as 90 percent for geothermal and 75 percent for cogeneration.

Chlorofluorocarbons (CFCs) A family of artificially produced chemicals that have been found to cause stratospheric ozone depletion. These molecules also power greenhouse gases. Since they were introduced in the mid-1930s, CFCs have been used as refrigerants, solvents, and in the production of foam material. The 1987 Montreal Protocol on CFCs seeks to reduce their production by one-half by the year 1998.

Cogeneration Production of heat energy and electrical or mechanical power from the same fuel in the same facility. A typical cogeneration facility produces electricity and steam for industrial process use.

Combined Cycle Plant An electric generating station that uses waste heat from its gas turbines to produce steam for conventional steam turbines.

Conservation Steps taken to cause less energy to be used than would otherwise be the case. These steps may involve improved efficiency, avoidance of waste, or reduced consumption. They may involve installing equipment, modifying equipment, adding insulation or other building modifications, or changing behavior.

Customer Class Utility customers are identified with a group that has several characteristics in common. Utilities typically classify customers as residential, commercial, industrial, and agricultural.

Decommissioning The dismantling of a power plant involving disassembling the equipment and transporting all material to a site for long-term disposal.

Decoupling A regulatory design that breaks the link between utility revenues and energy sales. This design is intended to align utility shareholders interests with ratepayer and societal interests.

Demand The level at which electricity is delivered to users at a given point in time. Electric demand is measured in kilowatts.

Demand-Side Management (DSM) The methods used to manage energy demand including energy efficiency, load management, fuel substitution, and load building.

Direct Energy Conversion Production of electricity from an energy source without first transferring the energy to a working fluid or steam. For example, photovoltaic cells transform light directly into electricity. Direct conversion systems have no moving parts and usually produce direct current.

Distribution System The substations, transformers, and lines that convey electricity from high-power transmission lines to ultimate consumers.

Economy Energy   Electricity a utility purchases that takes the place of electricity that would have cost more to produce on the utility's own system.

Electric Efficiency Ratio (EER)   The ratio of cooling capacity of an air conditioning unit in BTUs per hour to the total electrical input in watts under specified test conditions.

Electric Generator   A device that converts heat, chemical, or mechanical energy into electricity.

Electricity   A property of the basic parts of matter. A form of energy having magnetic, radiant, and chemical effects. A current of electricity is created by a flow of charged particles.

Energy   The capability of doing work. The resources that make a technology operational. The term "energy" is also used to mean electricity supplied over time. It is expressed in kilowatt-hours.

Energy Reserves   The portion of total energy resources that is known and can be recovered with presently available technology at an affordable cost.

Exchange   Agreements between utilities providing for purchase, sale, and trading of power. Usually relates to capacity (kilowatts) but sometimes energy (kilowatt-hours).

Externalities   Any costs or benefits not accounted for in the price of goods or services.

Federal Energy Regulatory Commission (FERC)   An independent federal commission that has jurisdiction over energy producers that sell or transport fuels for resale in interstate commerce.

Firm Energy   Power supplies that are guaranteed to be delivered under terms defined by contract.

Fission   A release of energy caused by the splitting of an atom's nucleus. This is the energy process used in conventional nuclear power plants to make the heat needed to run steam turbines.

Fissionable Material   A substance whose atoms can be split by slow neutrons. Uranium-235, plutonium-239, and uranium-233 are fissionable materials.

Fixed Cost   Costs of generation projects that must be paid regardless of the amount of energy produced. Such costs normally include capital costs, interest, insurance, and taxes.

Fluorescent Lamp   A electric lamp that is coated on its inner surface with a phosphor that when energized emits light of a selected color.

Fossil Fuel   Oil, coal, or natural gas. Fuel that was formed in the earth in prehistoric times from remains of living organisms.

Fuel Cell   A device that converts the chemical energy of fuel directly into electricity. The fuel cell does not burn the fuel and does not produce steam. It uses an electrical process that causes hydrogen atoms to give up their electrons.

Gasification The process used to produce power from biomass gas containing hydrogen, methane, carbon monoxide, nitrogen, water, and carbon dioxide. The gas can be burned directly in a boiler or scrubbed and combusted in an engine-generator to produce electricity. *Gasification* is also the production of synthetic gas from coal.

Geothermal Energy Natural heat from within the earth, captured for production of electric power, space heating, or industrial steam.

Gigawatt One thousand megawatts (1,000 MW) or one million kilowatts (1,000,000 kW) or one billion watts.

Global Climate Change Gradual changing of global climates due to buildup of carbon dioxide and other greenhouse gases in the earth's atmosphere. Carbon dioxide produced by burning fossil fuels has reached levels greater than what can be absorbed by plants and the seas.

Greenhouse Effect The process of trapping of longwave radiation by atmospheric gases, which makes the earth warmer than would direct sunlight alone. These gases (carbon dioxide, methane, nitrous oxide, tropospheric ozone, and water vapor) allow visible light and ultraviolet light (shortwave radiation) to pass through the atmosphere and heat the earth's surface. The heat is re-radiated from the earth in the form of infrared energy (longwave radiation). The greenhouse gases absorb part of that energy before it escapes into space.

Grid The electric transmission and distribution system that links power plants to customers.

Heat Rate A number that indicates the efficiency of a fuel-burning power plant. The heat rate equals the Btu content of the fuel input divided by the kilowatt hours of output.

Hydroelectric Power Electricity produced by falling water that turns a turbine generator. Sometimes it is referred as hydro.

Incandescent Lamp An electric lamp in which a filament is heated by an electric current until it emits visible light.

Interchange The agreement among interconnected utilities under which they buy, sell, and exchange power among themselves. This can provide for economy energy and emergency power supplies.

Interconnection The linkage of transmission lines between two utilities, enabling power to be moved in either direction. Interconnections allow utilities to enhance system reliability.

Interruptible Service Electricity supplied under agreements that allow the supplier to curtail or stop service at times.

Intertie A transmission line that links two or more regional electric power systems.

Landfill Gas Gas generated by the natural decomposition of municipal solid waste by anaerobic microorganisms in landfills. The gases produced, carbon dioxide and methane, can be collected and processed into a medium-Btu gas that can be burned to generate steam or electricity.

Levelized Life-Cycle Costs   The present value of a resource's cost converted to a stream of equal annual payments. This stream of payments can be converted to a unit cost of energy by dividing by the number of kilowatt-hours produced or saved by the resource. By levelizing costs, resources with different lifetimes and generating capabilities can be compared to one another.

Load   The amount of electric power supplied to meet one or more end user's needs.

Load Diversity   The condition that exists when the peak demands of a variety of electric customers occur at different times.

Load Factor   A percentage specifying the difference between the amount of electricity a consumer used during a given time span and the amount that would have been used if the usage had stayed at the consumer's higher demand level during the entire period of time.

Load Management   Steps taken to reduce power demand at peak load times or to shift some of the load to off-peak times. The main appliance affecting electric peaks is air conditioning, which is often a target for load management.

Losses   Electricity that is wasted in the normal operation of a power system. Some electricity is lost in the form of waste heat in electrical apparatus such as substation conductors.

Lumen   A measure of the amount of light available from a light source.

Lumens/Watt   A measure of the efficiency of a light fixture. The number of lumens output per watt of power input.

Marginal Cost   The amount that has to be paid for the next increment of product or service. The marginal cost of electricity is the price to be paid for kilowatt-hours beyond those supplied by the presently available generating capacity.

Methane   A light hydrocarbon that is the main component of natural gas.

Natural Gas   Hydrocarbon gas found in the earth, composed of methane, ethane, butane, propane, and other gases.

Nominal Dollars   Dollars that include the effects of inflation (*see* Real Dollars)

Non-Firm Energy   Electricity that is not required to be delivered or to be taken under the terms of an electric purchase contract.

NO$_x$ Oxides of nitrogen that are a chief component of air pollution and that can be produced by the burning of fossil fuels.

Nuclear Energy   Power obtained by splitting heavy atoms (fission) or joining light atoms (fusion). A nuclear power plant uses a controlled atomic chain reaction to produce heat. The heat is used to make steam to run conventional turbine generators.

Nuclear Regulatory Commission (NRC)   An independent federal agency that ensures that standards of public health and safety, environmental quality, and national security are adhered to by individuals and agencies possessing and using radioactive materials. The NRC is the agency that licenses and regulates nuclear power plants.

Outage   An interruption of electric service that is temporary and affects a relatively small area.

Ozone   A kind of oxygen that has three atoms per molecule instead of the usual two. Ozone is a poisonous gas, but the ozone layer in the upper atmosphere shields life on earth from damaging ultraviolet radiation from the sun.

Particulate Matter   Solid particles that are released from combustion processes in exhaust gases at fossil-fuel power plants.

Passive Solar Energy   Use of the sun to help meet a building's energy needs by means of architectural design or materials.

Peak Load   The highest electrical demand within a particular period of time. In warm weather climates electric peaks occur on weekdays during the late afternoon and early evening. Annual peaks occur on hot summer days.

Peaking Unit   A power plant used by a utility to produce electricity during peak load periods.

Photovoltaic Cell   A semiconductor that converts light directly into electricity.

Power Pool   Two or more interconnected utilities that plan and operate to supply electricity in the most reliable, economical way to meet their combined load.

Pressurized Water Reactor (PWR)   A nuclear power plant cooled by water that is pressurized to keep it from boiling when it reaches high temperatures.

PURPA   The Public Utilities Regulatory Policies Act of 1978. Under PURPA electric utilities are required to purchase available electricity from cogeneration plants and small power plants.

Qualifying Facility   A cogenerator or small power plant that under PURPA has the right to sell power to an electric utility.

Rankine Cycle   The Rankine cycle steam turbine has been the mainstay of utility power plants for many years. Rankine systems operate at a top temperature of about 1,073°C with maximum efficiencies of about 40 percent.

Reactor   A device in which a controlled nuclear chain reaction can be maintained, producing heat.

Real Dollars   Dollars that are adjusted to net out the effects of inflation.

Renewable Energy   Resources that constantly renew themselves or that are regarded as practically inexhaustible. These include solar, wind, geothermal, hydro, and biomass. Although geothermal formations can be depleted, the natural heat of the earth is a virtually inexhaustible reserve of potential energy. Renewable energy can also come from tidal power, sea currents, and ocean thermal gradients.

Reserve Margin   The extra generating capacity that an electric utility needs, above and beyond the highest demand level it is required to supply to meet its user's needs.

Retail Wheeling   The transmission of power from competing suppliers over a utility's power lines directly to retail customers.

Retrofit   Adding equipment to a facility or building after construction has been completed. Adding equipment tends to cost more than installing the equipment as part of the original construction.

Solar Power   Electricity generated by solar radiation.

Solar Radiation   The amount of radiation, both direct and diffuse, that can be received at any given location.

Solar Thermal   The process of concentrating sunlight on a relatively small area to create the high temperatures needed to vaporize water or other fluids to drive a turbine for the generation of electric power.

Stirling Engine   An external combustion engine that converts heat into useable mechanical energy by the heating and cooling of a captive gas.

Substation   A facility that steps up or steps down the voltage in utility power lines. Voltage is stepped up when power is sent through long-distance transmission lines. It is stepped down when the power enters local distribution lines.

Surplus   Excess firm energy available from a utility or region.

Thermal Energy Storage   A technology that lowers the amount of electricity needed for comfort conditioning during utility peak load periods.

Time-of-Use Rates   Electricity prices that vary depending on the time periods in which the energy is consumed. Higher prices are charged during utility peak load times. Such rates provide an incentive for consumers to curb power use during peak times.

Transmission   Transporting bulk power over long distances.

Turbine Generator   A device that uses steam, heated gases, water flow, or wind to cause a spinning motion that activates electromagnetic forces and generates electricity.

Vertical Integration   The control or ownership of all the different aspects of making, selling, and delivering a product or service by a single company.

Wheeling   Using a utility's power lines to transport electricity from one electric system to another.

Wholesale Competition   A system in which a distributor of electricity has the option to buy bulk power from a variety of power producers and the power producers are able to compete to sell electricity to a variety of distribution utilities.

Resource Guide

Alliance for Affordable Energy
604 Julia Street
New Orleans, LA 70130
phone   (504) 525-0778
fax   (504) 525-0779

American Planning Association
707 Park Avenue East
Tallahassee, FL 32301
phone   (904) 222-0808
fax   (904) 222-3741

American Wind Energy Association
122 C Street, NW, Fourth Floor
Washington, DC 20001
phone   (202) 383-2510
fax   (202) 383-2505

Businesses for Social Responsibility
Environmental and Energy Program
1030 15th Street, NW
Suite 1010
Washington, DC 20005
phone   (202) 842-5400
fax   (202) 842-3135

Campaign for a Prosperous Georgia
1083 Austin Avenue, NE
Atlanta, GA 30307
phone   (404) 659-5675
fax   (404) 659-5676

Center for Clean Air Policy
444 North Capitol Street, Suite 602
Washington, DC 20001
phone   (202) 624-7709
fax   (202) 508-3829

Center for Energy Efficiency and
Renewable Technologies, Research
and Education Fund
1100 11th Street, Suite 321
Sacramento, CA 95814
phone   (916) 442-7785
fax   (916) 447-2940

Citizen Fund
1120 19th Street, NW, Suite 630A
Washington, DC 20036
phone   (202) 332-0900
fax   (202) 332-0905

Citizens Action Coalition of Indiana
309 West Washington Street
Suite 233
Indianapolis, IN 46204
phone (317) 636-5165
fax (317) 636-5435

Citizens Energy Coalition
Education Fund
3951 North Meridian Street
Suite 300
Indianapolis, IN 46208
phone (317) 921-1120
fax (317) 921-1143

Conservation Law Foundation
62 Summer Street
Boston, MA 02110
phone (617) 350-0990
fax (617) 350-4030

Environmental Defense Fund
5655 College Avenue
Oakland, CA 94618
phone (510) 658-8008
fax (510) 658-0630

Environmental Defense Fund
Texas Office
44 East Avenue, Suite 304
Austin, TX 78701
phone (512) 478-5161
fax (512) 478-8140

Environmental Law and Policy
Center of the Midwest
203 North LaSalle Street, Suite 1390
Chicago, IL 60601
phone (312) 759-3400
fax (312) 332-1580

Izaak Walton League of America
Midwest Office
5701 Normandal Boulevard
Suite 210
Minneapolis, MN 55424
phone (612) 922-1608
fax (612) 922-0240

Land and Water Fund of the Rockies
2260 Baseline Road, Suite 200
Boulder, CO 80302-7740
phone (303) 444-1188
fax (303) 786-8054
email landwater@igc.apc.org

Lawrence Berkeley Laboratory
Energy Analysis Program
University of California, MS 90/400
Berkeley, CA 94720
phone (510) 486-5238
fax (510) 486-6996

Legal Environmental Assistance
Foundation
Energy Advocacy Program Director
1115 North Gadsden Street
Tallahassee, FL 32303-6327
phone (904) 681-2591
fax (904) 224-1275

Legislative Assistance Project
7870 Olentangy River Road
Suite 209
Columbus, OH 43235
phone (614) 888-7785
fax (614) 888-9716

Massachusetts PIRG
Education Fund
29 Temple Place
Boston, MA 02111
phone (617) 292-4800
fax (617) 292-8057

Michigan United Conservation Clubs
2101 Wood Street
P.O. Box 30235
Lansing, MI 48909
phone (517) 371-1041
fax (517) 371-1505

Mid Atlantic Energy Project
21 White Birch Drive
Pomona, NY 10970-3403
phone (503) 233-4544
fax (503) 223-4554

Minnesotans for an
Energy-Efficient Economy
Minnesota Building
46 East Fourth Street, Suite 1106
St. Paul, MN 55101
phone (612) 225-1133
fax (612) 225-0870

National Consumer Law Center, Inc.
18 Tremont Street, Suite 400
Boston, MA 02108-2336
phone (617) 523-8010
fax (617) 523-7398

Natural Resources Defense Council
Project for Sustainable FERC
Energy Policy
1350 New York Avenue, NW
Suite 300
Washington, DC 20005
phone (202) 783-7800
fax (202) 783-5917

Natural Resources Defense Council
71 Stevenson Street, Suite 1825
San Francisco, CA 94105
phone (415) 777-0220
fax (415) 495-5996

Natural Resources Defense Council
40 West 20th Street
New York, NY 10011
phone (212) 727-2700
fax (212) 727-1773

New Jersey PIRG
11 North Willow Street
Trenton, NJ 08608
phone (609) 394-8155
fax (609) 989-9013

Northeast Midwest Institute
218 D Street, SE
Washington, DC 20003
phone (202) 544-5200
fax (202) 544-0043

Northwest Conservation
Act Coalition
217 Pine Street, Suite 1020
Seattle, WA 98101-1520
phone (206) 621-0094
fax (206) 621-0097

Oak Ridge National Laboratory
P.O. Box 2008
Oak Ridge, TN 37831-6206
phone (615) 574-6304
fax (615) 576-8745

Ohio Environmental Council
Campaign for an Energy
Efficient Ohio
400 Dublin Avenue, Suite 120
Columbus, OH 43215-2333
phone (614) 224-4900
fax (614) 224-4914

Pace University
78 North Broadway
White Plains, NY 10603
phone (914) 422-4386
fax (914) 422-4180

Pace University Law School
Energy Project
Center for Environmental Studies
78 North Broadway
White Plains, NY 10603-3796
phone (914) 422-4141
fax (914) 422-4180

Public Citizen
Texas Branch
1800 Rio Grande
Austin, TX 78701
phone (512) 477-1155
fax (512) 479-8302

Regulatory Assistance Project
177 Water Street
Gardiner, ME 04345
phone (207) 582-1135
fax (207) 582-1176

Renew Wisconsin
222 South Hamilton
Madison, WI 53703
phone (608) 255-4044
fax (608) 251-7870

Renewable Northwest Project
1130 SW Morrison, Suite 330
Portland, OR 97205
phone (503) 223-4544
fax (503) 223-4554

Safe Energy Communication Council
1717 Massachusetts Avenue, NW
Suite 805
Washington, DC 20036
phone (202) 483-8491
fax (202) 234-9194
email seccgen@aol.com

Southern Environmental Law Center
201 West Main Street, Suite 14
Charlottesville, VA 22901
phone (804) 977-4090
fax (804) 977-1483

Sustainability Initiatives
2260 Baseline Road, Suite 200
Boulder, CO 80302-7740
phone (303) 417-1350
fax (303) 417-1351

Tennessee Valley Energy
Reform Coalition
P.O. Box 1842
Knoxville, TN 37901-1842
phone (615) 637-6055
fax (615) 524-4479

Texas Citizen Action
1714 Fortview, Suite 103
Austin, TX 78704-7659
phone (512) 444-8588
fax (512) 444-3533

Union of Concerned Scientists
Climate Change and Energy
Policy Program
1616 P Street, NW, Suite 310
Washington, DC 20036
phone (202) 332-0900
fax (202) 332-0905

Union of Concerned Scientists
2 Brattle Square, 6th Floor
Cambridge, MA 02238
phone (617) 547-5552
fax (617) 864-9405

University of Maryland
at College Park
Center for Global Change
7100 Baltimore Avenue, Suite 304
College Park, MD 20740
phone (301) 403-4165
fax (301) 403-4292

Widener University
3800 Verten Way
Harrisburg, PA 17110-8450
phone (717) 541-1967
fax (717) 541-1970

Wisconsin's Environmental
Decade Institute
122 State Street, #200
Madison, WI 53703
phone (608) 251-7020
fax (608) 251-1655

Bibliography

Bedford, Henry F. 1990. *Seabrook Station: Citizen Politics and Nuclear Power.* Amherst: The University of Massachusetts Press.

Brower, Michael, Michael Tennis, Eric Denzler, and Mark Kaplan. 1993. *Powering the Midwest: Renewable Electricity for the Economy and the Environment.* Cambridge, MA: Union of Concerned Scientists.

Cohen, Armond. 1992. *Power to Spare II: Energy Efficiency and New England's Economic Recovery.* Boston: Conservation Law Foundation.

Division of Ratepayer Advocates. 1994. *Estimates of Competitive Transition Costs or Uneconomic Costs and Obligations.* San Francisco: California Public Utilities Commission.

Division of Strategic Planning. 1993. *California's Electric Services Industry: Perspectives on the Past, Strategies for the Future.* San Francisco: California Public Utilities Commission.

Edson + Modisette. 1993. *Independent Power: A California Success Story.* Sacramento: Independent Energy Producers Association.

Feldman, Marvin, and Richard McCann. 1995. *The Effects of California Electricity Market Restructuring on Emerging Technologies.* Sacramento: California Energy Commission.

Flavin, Christopher, and Nicholas Lenssen. 1994. *Power Surge: Guide to the Coming Energy Revolution.* New York: W.W. Norton and Company.

Gipe, Paul. 1995. *Wind Energy Comes of Age.* New York: John Wiley & Sons.

Hamrin, Jan, and Nancy Rader. 1993. *Investing in the Future: A Regulator's Guide to Renewables.* Washington, DC: National Association of Regulatory Utility Commissioners.

221

Kahn, Edward. 1988. *Electric Utility Planning and Regulation*. Washington, DC: American Council for an Energy-Efficient Economy.

Kahrl, William L. 1982. *Water and Power*. Berkeley: University of California Press.

Kirlin, John. 1992. *California Policy Choices*, Vol. 8. Sacramento: University of Southern California.

Krause, Florentin, and Joseph Eto. 1988. *Least-Cost Utility Planning Handbook for Public Utility Commissioners*. Washington, DC: National Association of Regulatory Utility Commissioners.

Leigland, James and Robert Lamb. 1986. *WPP$$: Who is to Blame for the WPPSS Disaster?* Cambridge, MA: Ballinger Publishing Company.

Lovins, Amory B. and L. Hunter Lovins. 1982. *Brittle Power: Energy Strategy for National Security*. Amherst, NH: Brick House Publishing Company.

Mazuzan, George T., and J. Samuel Walker. 1984. *Controlling the Atom: The Beginnings of Nuclear Regulation 1946–1962*. Berkeley: University of California Press.

McCraw, Thomas K. 1984. *Prophets of Regulation*. Cambridge, MA: The Belknap Press of Harvard University Press.

McPhee, John. 1974. *The Curve of Binding Energy*. New York: The Noonday Press.

Moskowitz, David. 1992. *Renewable Energy: Barriers and Opportunities; Walls and Bridges*. Gardiner, ME: Regulatory Assistance Project.

Navarro, Peter. 1985. *The Dimming of America: The Real Costs of Electric Utility Regulatory Failure*. Cambridge, MA: Ballinger Publishing Company.

New England Energy Planning Council. 1987. *Power to Spare: A Plan for Increasing New England's Competitiveness Through Energy Efficiency*. Boston: New England Energy Policy Council.

Northwest Conservation Act Coalition. 1993. *Plugging People into Power: An Energy Participation Handbook*. Seattle: Northwest Conservation Act Coalition.

Nye, David. 1990. *Electrifying America: Social Meaning of a New Technology*. Cambridge, MA: The MIT Press.

Ogden, Douglas. 1996. *Boosting Prosperity: Reducing the Threat of Global Climate Change Through Sustainable Energy Investments*. Washington, DC: Environment Information Center.

Riccio, Jim and Grynberg, Michael. 1995. *A Roll of the Dice: NRC's Efforts to Renew Nuclear Reactor Licenses*. Washington, DC: Public Citizen.

Rudolph, Richard, and Scott Ridley. 1986. *Power Struggle: The Hundred-Year War Over Electricity*. New York: Harper and Row.

Sacramento Municipal Utility District. 1994. *SMUD Solar Program Plan: Making the Most of Solar Opportunities.* Sacramento: Sacramento Municipal Utility District.

Sacramento Municipal Utility District. 1995. *1995 Integrated Resource Plan: Achieving Municipal Power Goals in a Competitive Age.* Sacramento: Sacramento Municipal Utility District.

Sacramento Municipal Utility District. 1995. *Advanced and Renewable Technologies Development Program.* Sacramento: Sacramento Municipal Utility District.

Sacramento Municipal Utility District. 1995. *Benefits of the Urban Forest: A Case History of Sacramento's Shade Tree Program.* Sacramento: Sacramento Municipal Utility District.

Sacramento Municipal Utility District. 1995. *Demand-side Management Strategic Plan 1996–2000: Striking a Competitive Balance.* Sacramento: Sacramento Municipal Utility District.

Stutz, John, Bruce Biewald, and Daljit Singh. 1996. *Can We Get There from Here?—The Challenge of Restructuring the Electricity Industry So That All Can Benefit.* Boston: The Tellus Institute.

Tennessee Valley Authority. 1995. *Energy Vision 2020.* Knoxville: Tennessee Valley Authority.

Thayer, Robert. 1994. *Gray World, Green Heart: Technology, Nature and the Sustainable Landscape.* New York: John Wiley & Sons.

Ward, Ruth Sutherland. 1973. *". . . For the People"—The Story of the Sacramento Municipal Utility District.* Sacramento: Sacramento Municipal Utility District.

Index

225

Station
historical background, 21
market power of, 98
percentage of revenues for energy efficiency programs, 55
restructuring of, 88, 92, 93, 171, 172
stranded costs of, 95
San Onofre Nuclear Generating Station
(SONGS), 22, 68, 95–96
electricity market and, 82, 87, 172
marine mitigation plan, 96
settlement, 94–97
stranded costs of, 107 n1, 154
SCE. *See* Southern California Edison
Schwent, Vince, 197
Seabrook nuclear plants, 35, 38, 46, 110,
111, 170 n1
Seattle City Light, 55
Selin, Ivan, 152
Shaad, Paul, 14
Shimshak, Rachel, 124
Shoreham nuclear plant, 38
Siemens, 181, 182
Sierra Club, 86, 106, 198
Smart energy management systems, xii
Smith, Steve, 131
Smith, Tom, 133
Smog. *See* Air pollution
SMUD. *See* Sacramento Municipal Utility
District
Snohomish County Public Utility District,
202
Solar dish systems, 183
Solar power, 18, 49, 66, 215
competitiveness of, 103
cost of, 72
percentage of California electricity from,
22
for water heating, 176–177
Solar radiation, 215
Solar thermal, 182–184, 190, 215
Solid waste-to-energy facilities, 195
SONGS. *See* San Onofre Nuclear Generating Station
Southern California Edison (SCE). *See also*
Palo Verde nuclear plant; San
Onofre Nuclear Generating Station
breakup of monopoly, 88, 92, 93, 171
BRPU competitive bidding debate, 70–71
historical background, 10, 16, 21, 22
market power of, 97–98
percentage of revenues in energy efficiency programs, 1991, 55
role of federal government, 140
sale of power to SMUD, 35, 37, 42, 48

stranded costs of, 95
Southern Environmental Law Center, 220
South Texas nuclear plant, 107 n1, 132,
135
Soviet Union, former, 28–30, 149–150
Spot energy market, 85, 89–90, 168
State government. *See also individual state
names*; New England; Pacific Northwest
jurisdictional overlap with federal government, 2, 141
regulation of investor-owned utilities,
10–11, 16, 22, 67–69
role in deregulation, 2, 109
Steamboat Springs, 188
Steam host, 20
Steam turbines, 13–14, 19, 20, 180, 181
Stirling engine, 183, 215
Stone & Webster, 46
Stranded assets, 151, 155
Stranded benefits, 98–102
Stranded costs, 151, 154–155, 166
California, 77–78, 80–83, 90, 93–94, 95,
104–105
FERC proposed policy for recovery,
141–142
largest single source of, 107 n1
Municipal utilities in California, 95
national distribution of liability, 107 n2
New England, 116–117
Pacific Gas & Electric, 95
Pacific Northwest, 126
proposal to waive, 198
from Rancho Seco nuclear plant, 83
San Diego Gas & Electric, 95
from San Onofre Nuclear Generating Station, 107 n1, 154
Southern California Edison, 95
Tennessee Valley Authority, 128–131
Texas, 137
utility executive poll, 170 n2
Strumwasser, Michael, 82
Students for a Democratic Society, 35
Subsidies, 150–152, 172
Substations, 20, 92, 215
Sulfur dioxide emissions, 2–3, 74 n3, 103,
149
Surplus electricity, 215. *See also* Electricity
markets; Wholesale power pools
Surveys, on conservation and renewable
energy, 199–206
Sustainability Initiatives, 220
Sustainable Energy Budget Coalition, survey
results, 199–200, 202–203
Systems benefit charge, 102, 105, 107, 144